APPLICATIONS OF NUMERICAL HEAT TRANSFER

SERIES IN THERMAL AND FLUIDS ENGINEERING

JAMES P. HARTNETT and THOMAS F. IRVINE, JR., Editors
JACK P. HOLMAN, Senior Consulting Editor

Cebeci and Bradshaw • Momentum Transfer in Boundary Layers
Chang • Control of Flow Separation: Energy Conservation, Operational Efficiency, and Safety
Chi • Heat Pipe Theory and Practice: A Sourcebook
Eckert and Goldstein • Measurements in Heat Transfer, 2d edition
Edwards, Denny, and Mills • Transfer Processes: An Introduction to Diffusion, Convection, and Radiation
Fitch and Surjaatmadja • Introduction to Fluid Logic
Ginoux • Two-Phase Flows and Heat Transfer with Application to Nuclear Reactor Design Problems
Hsu and Graham • Transport Processes in Boiling and Two-Phase Systems, Including Near-Critical Fluids
Kestin • A Course in Thermodynamics, revised printing
Kreith and Kreider • Principles of Solar Engineering
Lu • Introduction to the Mechanics of Viscous Fluids
Moore and Sieverding • Two-Phase Steam Flow in Turbines and Separators: Theory, Instrumentation, Engineering
Nogotov • Applications of Numerical Heat Transfer
Richards • Measurement of Unsteady Fluid Dynamic Phenomena
Sparrow and Cess • Radiation Heat Transfer, augmented edition
Tien and Lienhard • Statistical Thermodynamics, revised printing
Wirz and Smolderen • Numerical Methods in Fluid Dynamics

PROCEEDINGS

Hoogendoorn and Afgan • Energy Conservation in Heating, Cooling, and Ventilating Buildings: Heat and Mass Transfer Techniques and Alternatives
Keairns • Fluidization Technology
Spalding and Afgan • Heat Transfer and Turbulent Buoyant Convection: **Studies and Applications for Natural Environment, Buildings, Engineering Systems**
Zarić • Thermal Effluent Disposal from Power Generation

APPLICATIONS OF NUMERICAL HEAT TRANSFER

E. F. Nogotov
Luikov Institute for Heat and Mass Transfer
Academy of Sciences, Byelorussian SSR

B. M. Berkovsky, *Editor*
Science Sector
UNESCO

W. J. Minkowycz, *Consulting Editor*
University of Illinois at Chicago Circle

unesco

Paris

HEMISPHERE PUBLISHING CORPORATION

Washington   London

McGRAW-HILL BOOK COMPANY

New York   St. Louis   San Francisco   Auckland   Bogotá
Düsseldorf   Johannesburg   London   Madrid   Mexico
Montreal   New Delhi   Panama   Paris   São Paulo
Singapore   Sydney   Tokyo   Toronto

APPLICATIONS OF NUMERICAL HEAT TRANSFER

First published 1978 by the United Nations Educational, Scientific, and Cultural Organization, 7 Place de Fontenoy, 75700 Paris, France, Hemisphere Publishing Corporation, 1025 Vermont Avenue, N.W., Washington, D.C. 20005 U.S.A., and McGraw-Hill Book Company, 1221 Avenue of the Americas, New York, New York 10020 U.S.A.

© Unesco 1978. All rights reserved. Printed in the United States of America. No part of this publication may be reproduced, stored in a retrieval system, or transmitted, in any form or by any means, electronic, mechanical, photocopying, recording, or otherwise without the prior written permission of the publisher.

1 2 3 4 5 6 7 8 9 0   D O D O   7 8 3 2 1 0 9 8

This book was set in Press Roman by Hemisphere Publishing Corporation. The editors were Judith B. Gandy and Edward Millman; the production supervisor was Rebekah McKinney; and the compositor was Wayne Hutchins.
R. R. Donnelley & Sons Company was printer and binder.

Library of Congress Cataloging in Publication Data

Nogotov, Evgeniĭ Fomich.
  Applications of numerical heat transfer.

  Bibliography: p.
  Includes index.
  1. Heat—Transmission. 2. Nets (Mathematics).
I. Title.
TJ260.N64   1978   621.4'022   77-28006

Unesco: ISBN 92-3-101399-8
McGraw-Hill Book Company: ISBN 0-07-046852-4

Contents

Preface   vii

Chapter 1	**INTRODUCTION**	1
Chapter 2	**BASIC CONCEPTS OF THE FINITE-DIFFERENCE METHOD**	7

    2.1   Preliminary Remarks   7
    2.2   Approximation   12
    2.3   Stability; Convergence Theorem   16
    2.4   Stability Analysis of Finite-Difference Computations by the Fourier Transformation   22
    2.5   Stability Analysis of Finite-Difference Computations by the Maximum Principle   30
    2.6   Construction of Calculation Formulas   34
    2.7   Solution of Difference Equations   48

Chapter 3 HEAT–CONDUCTION PROBLEMS 57

- 3.1 Simple Heat-Conduction Problems 57
- 3.2 Finite-Difference Algorithms for One-Dimensional Heat-Conduction Problems with Constant Thermal Diffusivities 60
- 3.3 Characteristics of Numerical Solutions of Heat-Conduction Problems in Cylindrical and Spherical Geometries 67
- 3.4 Approximations of Boundary Conditions 69
- 3.5 Finite-Difference Procedures for Heat-Conduction Equations with Variable Coefficients 76
- 3.6 Fractional-Step Method 82
- 3.7 Two-Dimensional Heat-Conduction Problems 85
- 3.8 Difference Methods Applicable to Three-Dimensional Heat-Conduction Problems 91
- 3.9 Nonlinear Heat-Conduction Problems 74

Chapter 4 CONVECTIVE HEAT TRANSFER 99

- 4.1 Convection Equations; Boundary Conditions 99
- 4.2 Characteristics of Computation Algorithms 105
- 4.3 Numerical Procedures for Heat-Convection Procedures with Low and Medium Rates 109
- 4.4 High-Rate Convection Processes 116
- 4.5 Numerical Analysis of Steady-State Convective Problems 122
- 4.6 Convective Heat Transfer in Compressible Media 125

Chapter 5 CONCLUSION 131

- 5.1 General Recommendations for the Application of Finite-Difference Techniques to Heat-Transfer Problems 131

References 135
Index 139

Preface

Ideas for the practical application of numerical methods are finding an increasingly receptive audience among progressive engineers around the world. In the present work, numerical methods are applied to an important engineering problem, because it is believed that, in general, engineers will prefer a demonstration of this kind to an abstract mathematical presentation.

Engineers working in the energy field, in particular in the field of heat and mass transfer, will find in these pages the information they need for the employment of modern numerical methods in their daily practice. Most of the problems involved in heat transfer are reducible to the solution of partial differential equations. As a general rule, however, these are complicated and their solution in the form of final formulas is possible in only the simplest cases. The most important group of approximate methods of solution of these equations is composed of numerical ones and, of these, the finite-difference technique is certainly the most universal and widely used.

A complete and systematic treatment of the subject, from basic concepts of finite-difference methods to sophisticated finite-difference schemes, is provided in this work. General recommendations for the application of the finite-difference technique are given, as well as examples of the solution of particular problems in heat transfer.

This book has been written within the framework of Unesco's program in technological research and higher education. Unesco wishes to thank the Byelorussian National Commission for Unesco for having proposed the author, and Mr. Nogotov himself for having carried out this valuable work. The author, who is an expert with wide experience in the application of numerical methods to heat-transfer problems, is responsible for the choice and presentation of materials and the opinions expressed in this study.

APPLICATIONS OF NUMERICAL HEAT TRANSFER

Chapter 1

Introduction

Many problems involving heat and mass transfer are reducible to the solution of partial differential equations. The differential equations that govern real physical processes are generally of a very complicated nature, and their closed-form solution is possible only in the simplest cases.

Approximate methods therefore become very useful for the solution of such problems. The methods generally are divided into two categories. The first category covers those methods that allow an analytical expression, say, a part of a certain series, as the approximate solution of the problem. It should be emphasized that in most cases such a solution has a complicated structure, as it contains integrals, special functions, etc., and is hardly convenient.

The second category of approximate methods is composed of numerical techniques that allow the determination of a table of approximate values of the desired solution. In this category are such approaches as the finite-difference method, straight-line method, large-particle method, and Monte Carlo method.

Of the numerical methods, the finite-difference technique is certainly the most universal and most widely used. The essence of the method involves substitution, for the differential operators in the initial differential equations, of approximate values expressed in terms of differences of the functions at discrete points of the differencing grid. The substitution results in algebraic systems of equations with the function values at the grid points being the unknowns. The method has a universal appeal because of its generality and its relative simplicity of adaptation to a computer.

The straight-line method is closely associated with the finite-difference technique. In this method, the solution of the partial differential equation is sought along a certain straight-line family; the equation is reduced to a system of ordinary differential equations, which may often be solved by a finite-difference procedure.

Multidimensional equations of mathematical physics can often be solved with a large-particle method suggested recently by Harlow. In this method, the hydrodynamic equations are reduced to two simpler systems in each time step, on the basis of some weak approximation. The first system describes the interaction of hydrodynamic fields, with transfer effects neglected, and it is integrated by ordinary means using a fixed Eulerian grid. The second system describes transfer effects. For its solution, a simplified continuum model is used, with a set of particles substituted for each Eulerian mesh. The net balance of the mass, momentum, and energy of particles in a mesh is identical to that in a continuum. As soon as a particle that "carries" a certain mass crosses a mesh boundary in following its path, the mass, momentum, and energy of the particle are subtracted from the mesh it has left, and are added to the mesh that now contains the particle.

Harlow's system is based on the explicit solutions of the first and second stages and is generally assumed quite stable as a whole. No absolutely stable schemes for large-particle methods are available as yet, but significant progress may be expected sometime in the near future.

Recently, O. M. Belotserkovsky, Y. M. Davydov, N. N. Yanenko, and others have presented some modifications of the method that essentially reduce the velocity and pressure fluctuations and increase the stability.

It should be noted that the structure of the method is rather complicated. Moreover, computations based on the method always require a large amount of computer time and storage, thus making the range of application of the method rather narrow.

During the last two decades, another method, known as the *Monte Carlo* method, has been actively developed. This method is most effective when used with high-speed computers, since it requires a large number of statistical tests to reduce the mean-square error of the result.

Recently, the method has been improved by the use of conditional probabilities of the processes and statistical weights based on the solution of the conjugate equations. In particular cases, this approach reduces the error dispersion by about an order, thus also decreasing the computation time by an order. Much more work is still required in this regard.

The wide use of numerical methods for the different problems of mathematical physics, including those of heat transfer, is closely tied to the current development of computers. Modern high-speed computers allow the numerical investigation of many important practical problems that are formulated in the most general terms.

The strict formulation of a problem with a small number of assumptions makes numerical investigation comparable to the best physical experiments. Moreover, numerical experiments possess a number of advantages such as relative simplicity and low cost and provide the possibility of accounting for most of the effects involved in the process being considered. Furthermore, numerical experiments permit the variation of problem parameters and boundary conditions over wide ranges, and they provide complete information on the process under investigation, which is practically impossible under laboratory conditions.

The increased availability of computers and associated algorithms, which considerably simplify programming, has made computer techniques accessible to a large number of investigators, and has accelerated the use of mathematical techniques in science and technology. As a result, the number of workers requiring a knowledge of the fundamentals of mathematics is rapidly increasing. For these workers, the basic methods of computation must be presented in an accessible form. It is hoped that this book will serve that purpose.

In this book, some aspects of the development and the application of numerical methods to heat- and mass-transfer problems are considered. Methods are limited to those that provide computer solutions to the types of problems most frequently encountered in engineering practice.

Since, for the practical engineer, knowledge of the solution algorithm and its accuracy is of primary importance, strict mathematical proofs are sometimes omitted here; the essence of a particular method and the procedure for estimating the error of the solution are presented only in their final form. The types of problems for which the method is applicable and the process parameters for which the method gives a meaningful solution are generally outlined.

The reader who wishes to become expert in computer work must carefully study the literature, which gives a deeper insight into the fundamentals of computational mathematics. A list of references is provided, for that purpose, at the end of the book. It is hoped that the mathematical methods contained in the book will allow the successful solution of a wide range of heat-transfer problems.

It is expected that readers will have different specialities and levels of mathematical knowledge. Therefore, the presentation will not follow the very formal style traditional for mathematical literature; nor will all the material be given on a "physical" level. The appeal will be to the engineering intuition and common sense, rather than to mathematical education.

Since heat-transfer problems are mainly formulated in terms of partial differential equations, and finite-difference techniques comprise the most convenient and general method for solving such problems, a large part of the book is devoted to the description of these techniques.

A considerable amount of experience has been accumulated on the solution of heat-transfer problems by finite-difference techniques. Unfortunately, the descriptions of particular methods are contained in many different journals, or in special literature concerned with numerical methods, where they are presented in a form hardly accessible to nonspecialists in numerical analysis. This book thus presents, in a convenient form, the important numerical algorithms for the solution of heat-transfer problems.

It is not the author's aim to describe all the available finite-difference schemes, a task that is almost impossible. The book contains only the simplest and most effective finite-difference techniques applicable to typical heat- and mass-transfer problems. In some cases, effective but complicated methods have been omitted.

To help the reader choose and apply particular finite-difference schemes, Chap. 2 presents some basic concepts of the finite-difference technique. Details are given on how to select a particular differencing grid and how to construct the corresponding finite-difference formulas. Definitions of approximation, stability, and convergence are also given.

Chapter 3 is devoted entirely to the solution of heat-conduction problems. It contains various finite-difference schemes for the solution of heat-conduction problems and, in addition, a detailed description of the fractional-step method for solving multidimensional problems. The most interesting modifications of this method are presented.

In Chap. 4, numerical procedures for studying convective heat transfer are considered. Recommended finite-difference schemes, which are applicable in particular to high Rayleigh numbers, are presented.

Chapter 5 gives general recommendations concerning the choice of a finite-difference scheme to solve a particular problem, arrangement for the computation procedure, experimental determination of the stability of a difference problem and the accuracy of the resultant solution, and how to determine local and integral characteristics of the problem.

The book is based on lectures delivered by the author at the International School on Mathematical Aspects of Heat and Mass Transfer during the conference on Heat and Mass Transfer in 1974 (Minsk, U.S.S.R.), which was sponsored by the International Centre of Academies of Socialist Countries. The final text of the book was considerably improved through numerous discussions with colleagues, to whom the author is extremely grateful.

Chapter 2

Basic Concepts of the Finite-Difference Method

2.1 PRELIMINARY REMARKS

To be solved on a computer, a problem must be formulated numerically in terms of some suitable arithmetic operations. Generally, the problem to be solved is first formulated in terms of ordinary mathematical equations, functions, differential operators, etc. Therefore, computer solution first involves an approximation of the problem substitution in terms of numbers and arithmetic operations. For example, special functions are ordinarily expressed in terms of finite series; finite sums are substituted for integrals; and differential operators are usually approximated by difference relations.

The method of solution of differential problems in which differential operators are replaced by their approximate values expressed in terms of functions at individual discrete points is called the *finite-difference* method or, alternatively, the *grid* method.

The substitution reduces the problem to the solution of a set of algebraic equations. Although the set of equations may involve a considerable number of unknown quantities, its solution is mathematically easier than that of the original problem.

The essence of the finite-difference method will be illustrated with a simple example. The heat-conduction equation for a homogeneous thin rod of unit length ($0 \leqslant x \leqslant 1$) with internal distributed heat sources is to be solved:

$$\frac{\partial u}{\partial t} - \kappa \frac{\partial^2 u}{\partial x^2} = Q \tag{2.1}$$

where $u = u(x, t)$ is the temperature, $Q = Q(x, t)$ is the distribution function of the heat sources, $\kappa = \text{const} > 0$ is the thermal conductivity of the material, and t is the time.

Let the temperature distribution in the rod at some initial time $t = 0$ be

$$u(x, 0) = T(x) \tag{2.2}$$

and the temperatures at the ends of the rod be

$$u(0, t) = T_0(t) \quad u(1, t) = T_1(t) \tag{2.3}$$

We seek the temperature distribution $u(x, t)$ in the rod at any time $t > 0$.

For simplicity and for greater clarity, the problem will be written in an operator form as

$$Lu = f \tag{2.4}$$

with the notations

$$Lu \equiv \begin{cases} \dfrac{\partial u}{\partial t} - \kappa \dfrac{\partial^2 u}{\partial x^2} \\ u(x, 0) \\ u(0, t),\, u(1, t) \end{cases} \quad f \equiv \begin{cases} Q(x, t) \\ T(x) \\ T_0(t),\, T_1(t) \end{cases}$$

Here L is the differential operator, and f is the right-hand side. The differential problem (2.4) is defined in the region $0 \leqslant x \leqslant 1$, $t \geqslant 0$, which will be denoted by G. A continuous solution $u(x, t)$ in the region G is assumed to exist. To find the solution by the finite-difference method, the

BASIC CONCEPTS OF THE FINITE-DIFFERENCE METHOD

original differential problem should be written in terms of its finite-difference equivalent. Thus, it is necessary to replace the region G of continuously changing arguments with a discrete set of points, denoted by $G_{h\tau}$ and referred to as a *grid*. The points are called the *grid points*. The functions to be determined in the discrete set are referred to as the *grid functions*.

The locations of the grid points in the region $G_{h\tau}$ may be arbitrary; they depend on the specific properties of the problem to be solved. For simplicity, the grid $G_{h\tau}$ considered here (Fig. 1) is assumed to cover the whole region uniformly, and the grid points are defined by the coordinates (x_i, t^n):

$$x_i = ih \quad i = 0, 1, 2, \ldots, J$$
$$t^n = n\tau \quad n = 0, 1, 2, \ldots$$

Thus, two more parameters (h and τ) are involved; they are called the *grid steps*, and they describe the distribution density of the grid points with respect to the time and space variables, respectively. As τ or h decreases, the spacing between the corresponding grid points becomes smaller.

Instead of the continuous functions $u(x, t)$ and $f(x, t)$, we shall use appropriate numerical functions, which, when defined at grid points (x_i, t^n), will be denoted respectively by u_i^n and f_i^n. In some cases, an alternative form of the grid function will be used. Such a function will be

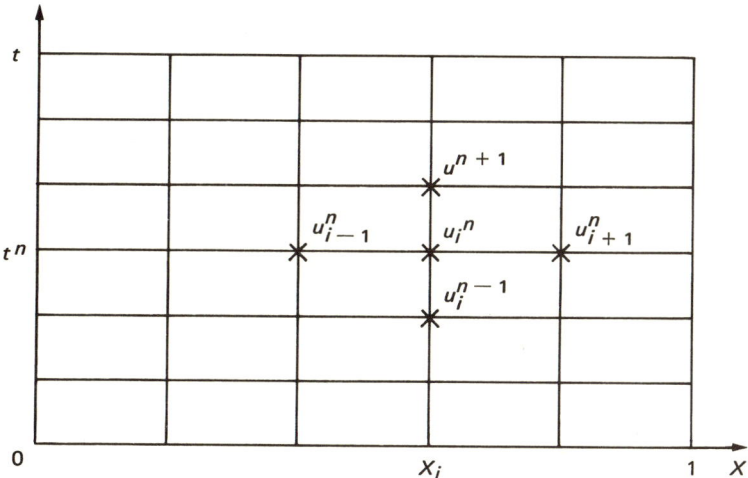

Fig. 1 The grid $G_{h\tau}$.

denoted by a letter with subscript h, for example, u_h and f_h. In the transition from a continuous function to a grid function, one ordinarily follows the rule that a value of the grid function at grid point (x_i, t^n) is the same as that of the respective continuous function at the same point:

$$u_i^n = u(x_i, t^n) \qquad f_i^n = f(x_i, t^n)$$

The above procedure is known as the *function projection* on the grid.

Different methods of determining grid functions are available. For example,

$$u_i^n = \frac{1}{h} \int_{x_i-h/2}^{x_i+h/2} u(x, t^n)\, dx \qquad f_i^n = \frac{1}{h} \int_{x_i-h/2}^{x_i+h/2} f(x, t^n)\, dx \qquad (2.5)$$

may be considered as u_i^n and f_i^n. This method is useful when $u(x, t)$ and $f(x, t)$ are not continuous functions and the integral of any of the functions over any finite interval is known to exist.

In what follows, unless otherwise specified, the values of grid functions will always be assumed to be the same as those of the continuous functions at the grid points.

So, the solution of the original differential problem (2.1)–(2.3) is the tabulation of numerical values u_i^n of the exact solution $u(x, t)$ at the grid points of $G_{h\tau}$. The calculation of these values requires the substitution, for the differential equation (2.1), of a difference equation approximating it at the grid points. The simplest way is to replace the derivatives of Eq. (2.1) with some difference relations. This replacement is based on the determination of the derivative as the limit

$$\frac{\partial u(x, t)}{\partial x} = \lim_{h \to 0} \frac{u(x + h, t) - u(x, t)}{h}$$

Indeed, if h is fixed in the above equality, the approximate formula for the first derivative expressed in terms of finite differences may be obtained:

$$\frac{\partial u(x, t)}{\partial x} \approx \frac{u(x + h, t) - u(x, t)}{h} \qquad (2.6)$$

The higher-order derivatives may be approximated in a similar way, e.g.,

BASIC CONCEPTS OF THE FINITE-DIFFERENCE METHOD

$$\frac{\partial^2 u(x,t)}{\partial x^2} \approx \frac{1}{h}\left[\frac{u(x+h,t)-u(x,t)}{h} - \frac{u(x,t)-u(x-h,t)}{h}\right]$$

$$= \frac{u(x+h,t) - 2u(x,t) + u(x-h,t)}{h^2} \tag{2.7}$$

The finite-difference equations involve grid functions. Relations (2.6) and (2.7) may be written for any point in the grid region $G_{h\tau}$ as

$$\left(\frac{\partial u}{\partial x}\right)_i^n \approx \frac{u_{i+1}^n - u_i^n}{h} \tag{2.8}$$

$$\left(\frac{\partial^2 u}{\partial x^2}\right)_i^n \approx \frac{1}{h^2}(u_{i+1}^n - 2u_i^n + u_{i-1}^n) \tag{2.9}$$

Replacement of the derivatives entering into equation (2.1) by their equivalent difference relations yields

$$\frac{U_i^{n+1} - U_i^n}{\tau} - \frac{\kappa}{h^2}(U_{i+1}^n - 2U_i^n + U_{i-1}^n) = Q$$

$$i = 1, 2, 3, \ldots, J-1 \quad n = 1, 2, 3, \ldots \tag{2.10}$$

The initial and boundary conditions expressed in a difference form are

$$\begin{aligned}U_i^0 &= T_i & i &= 1, 2, 3, \ldots, J-1 \\ U_0^n &= T_0^n \quad U_J^n = T_1^n & n &= 1, 2, 3, \ldots\end{aligned} \tag{2.11}$$

To summarize, the original differential problem (2.1)–(2.3) has been reduced to the difference problem (2.10)–(2.11). The above substitution for the derivatives is certainly not the only possible one, since many others are available. We shall, however, dwell upon the method. A combination of difference equations, together with initial and boundary conditions expressed in a difference form, will be referred to as a *finite-difference scheme*.

An algorithm for the solution of the problem (2.10)–(2.11) is very simple to develop. If the solution U_i^n for some time moment $t^n = n\tau$ is known, then the U_i^{n+1}'s may be calculated from the equation

$$U_i^{n+1} = \frac{\kappa\tau}{h^2}(U_{i+1}^n - U_{i-1}^n) + \left(1 - \frac{2\kappa\tau}{h^2}\right)U_i^n + \tau Q_i^n \tag{2.12}$$

Since, at the initial time $t = 0$, the U_i^0's are prescribed, we may easily determine $U_i^1, U_i^2, \ldots, U_i^n$, in succession.

The reader should have noticed that, in the formulation of the difference problem, U has been substituted for u, which denotes the exact solution of the original differential problem. This substitution emphasizes the difference between the solution U_i^n obtained from the difference problem and the projection on the grid points of the exact solution u_i^n obtained from the differential problem. The solutions are probably similar only in the case of small τ and h. However, it appears that even with vanishing τ and h, the solution of a difference problem does not always approach the exact solution of the initial problem. We shall dwell upon this point in more detail.

2.2 APPROXIMATION

After a differential problem has been replaced with its difference form, a question may naturally arise as to whether the solutions approach each other. Before the question is considered, it should be noted that, as with differential problems, an operator form of difference problems appears useful in some cases:

$$L_h U_h = f_h \tag{2.13}$$

where f_h is the grid function that is a projection of the right-hand side of the original differential problem $Lu = f$ on the grid, and L_h is the difference operator that approximates the differential operator L at the grid points. The operator L_h is determined from the grid functions and depends on the parameters τ and h of the grid. In particular, it is also determined from the grid function u_h, which is a projection of the exact solution $Lu = f$ on the grid. Thus, operation $L_h U_h$ and operation $L_h u_h$ produce a meaningful solution.

For example, the difference problem (2.10)-(2.11), when written in the operator form (2.13), becomes

$$L_h U_h \equiv \begin{cases} \dfrac{U_i^{n+1} - U_i^n}{\tau} - \dfrac{\kappa}{h^2}(U_{i+1}^n - 2U_i^n + U_{i-1}^n) \\ U_i^0 \\ U_0^n, U_J^n \end{cases} \qquad f_h \equiv \begin{cases} Q_i^n \\ T_i \\ T_0^n, T_1^n \end{cases}$$

$$\tag{2.14}$$

BASIC CONCEPTS OF THE FINITE-DIFFERENCE METHOD

Now we shall proceed to the main question, namely, whether the numerical solution U_i^n of difference problem (2.13) approaches the exact solution $u(x, t)$ of the original differential problem $Lu = f$. Since the exact solution $u(x, t)$ is a continuous function, while U_i^n is discrete, their comparison requires an extension of the solution U_i^n to the whole region $G_{h\tau}$ or, alternatively, a projection of the exact solution $u(x, t)$ on the grid points of $G_{h\tau}$. The latter procedure, being easier, is the one commonly used.

Let us compare the functions U_i^n and u_i^n (or, equivalently, U_h and u_h). The difference problem $L_h U_h = f_h$ is assumed to have a unique solution. Suppose, after the substitution of u_h for grid function U_h on the left-hand side of the above equality, it were exactly satisfied. Then, because of the uniqueness of the solution, the identity $U_h \equiv u_h$ would hold, which would imply that the solution of a difference problem is exactly the same as the grid function u_h, which is by convention the exact solution of the differential problem $Lu = f$. However, this situation is impossible. The substitution of u_h into (2.13) will always involve an additional term δf_h, that is,

$$L_h u_h = f_h + \delta f_h \qquad (2.15)$$

If, owing to the substitution of u_h (a projection of the exact solution u on the grid) into the difference problem, δf_h is vanishing as τ and h decrease, then we may say [1-3] that the difference problem $L_h U_h = f_h$ approximates the differential problem $Lu = f$ in the solution u.

Let $\tau, h \to 0$, along with $|\delta f_h| = |L_h u_h - f_h| \to 0$. If the inequality

$$|\delta f_h| \leqslant c_1 \tau^{k_1} + c_2 h^{k_2} \qquad (2.16)$$

holds where c_1, k_1 and c_2, k_2 are positive and independent of τ and h, then the difference problem is said to approximate the original differential problem with an order k_1 with respect to τ, and with an order k_2 with respect to h.

The study of the order of approximation is not usually difficult and is one and the same for all the problems; it is based on the Taylor-series expansion of the exact differential problem at mesh points. The determination of the approximation order of a differencing scheme will be illustrated by considering problem (2.10)-(2.11).

Let us therefore demonstrate that the difference scheme (2.10)-(2.11) approximates the differential problem (2.1)-(2.3) in its exact solution $u(x, t)$, and then determine the approximation order of the procedure. First, $\delta f_h = L_h u_h - f_h$ will be evaluated. Since, in the case under consideration, the boundary conditions are exactly approximated, the value of δf_h will be determined by the error of the approximation of initial equation (2.1) by the difference equation (2.10). Thus,

$$\delta f_h = \frac{u_i^{n+1} - u_i^n}{\tau} - \frac{\kappa}{h^2}(u_{i+1}^n - 2u_i^n + u_{i-1}^n) - Q_i^n \tag{2.17}$$

To evaluate δf_h, we shall expand the functions $u_i^{n+1} = u(x_i, t^n + \tau)$, $u_{i-1}^n = u(x_i - h, t^n)$, and $u_{i+1}^n = u(x_i + h, t^n)$ in the right-hand side of (2.17) as Taylor series in the vicinity of (x_i, t^n). It is necessary that the solution $u(x, t)$ satisfy the smoothness conditions and have a sufficient number of limited derivatives. Note that most of the transfer problems encountered in engineering and scientific practice have solutions that meet these requirements. We assume these requirements are satisfied and express the functions as

$$u(x_i, t^n + \tau) = u(x_i, t^n) + \tau \frac{\partial u(x_i, t^n)}{\partial t} + \frac{\tau^2}{2} \frac{\partial^2 u(x_i, \tilde{t})}{\partial t^2}$$

$$u(x_i + h, t^n) = u(x_i, t^n) + h \frac{\partial u(x_i, t^n)}{\partial x} + \frac{h^2}{2} \frac{\partial^2 u(x_i, t^n)}{\partial x^2} + \frac{h^3}{3!} \frac{\partial^3 u(x_i, t^n)}{\partial x^3}$$

$$+ \frac{h^4}{4!} \frac{\partial^4 u(\tilde{x}, t^n)}{\partial x^4}$$

$$u(x_i - h, t^n) = u(x_i, t^n) - h \frac{\partial u(x_i, t^n)}{\partial x} + \frac{h^2}{2} \frac{\partial^2 u(x_i, t^n)}{\partial x^2} - \frac{h^3}{3!} \frac{\partial^3 u(x_i, t^n)}{\partial x^3}$$

$$+ \frac{h^4}{4!} \frac{\partial^4 u(\tilde{\tilde{x}}, t^n)}{\partial x^4}$$

where $\tilde{t}$, $\tilde{x}$, and $\tilde{\tilde{x}}$ are some fixed values of the variables t and x satisfying the inequalities $t^n \leqslant \tilde{t} \leqslant t^{n+1}$, $x_i \leqslant \tilde{x} \leqslant x_{i+1}$, and $x_{i-1} \leqslant \tilde{\tilde{x}} \leqslant x_i$.

From these relations we may write

$$\frac{u_i^{n+1} - u_i^n}{\tau} = \left(\frac{\partial u}{\partial t}\right)_i^n + \frac{\tau}{2} \frac{\partial^2 u(x_i, \tilde{t})}{\partial t^2}$$

$$\frac{1}{h^2}(u_{i+1}^n - 2u_i^n + u_{i-1}^n) = \left(\frac{\partial^2 u}{\partial x^2}\right)_i^n + \frac{h^2}{24}\left[\frac{\partial^4 u(\tilde{x}, t^n)}{\partial x^4} + \frac{\partial^4 u(\tilde{\tilde{x}}, t^n)}{\partial x^4}\right]$$

$$\tag{2.18}$$

BASIC CONCEPTS OF THE FINITE-DIFFERENCE METHOD

Here the indices in the derivatives indicate that the derivatives are taken at (x_i, t^n). Substitution of the above expressions into (2.17) yields

$$\delta f_h = \left(\frac{\partial u}{\partial t}\right)_i^n - \kappa \left(\frac{\partial^2 u}{\partial x^2}\right)_i^n - Q_i^n + \frac{\tau}{2} \frac{\partial^2 u(x_i, \tilde{t})}{\partial t^2}$$

$$- \frac{\kappa h^2}{24} \left[\frac{\partial^4 u(\tilde{x}, t^n)}{\partial x^4} + \frac{\partial^4 u(\tilde{\tilde{x}}, t^n)}{\partial x^4}\right]$$

Since $u(x, t)$ exactly satisfies (2.1) and

$$\left(\frac{\partial u}{\partial t}\right)_i^n - \kappa \left(\frac{\partial^2 u}{\partial x^2}\right)_i^n = Q_i^n$$

then

$$\delta f_h = \frac{\tau}{2} \frac{\partial^2 u(x_i, \tilde{t})}{\partial t^2} - \frac{\kappa h^2}{24} \left[\frac{\partial^4 u(\tilde{x}, t^n)}{\partial x^4} + \frac{\partial^4 u(\tilde{\tilde{x}}, t^n)}{\partial x^4}\right] \qquad (2.19)$$

Since all the derivatives entering into inequality (2.19) are limited and have an exact upper boundary, the approximation will be

$$|\delta f_h| \leq \frac{\tau}{2} \sup \left|\frac{\partial^2 u}{\partial t^2}\right| + \frac{\kappa h^2}{12} \sup \left|\frac{\partial^4 u}{\partial x^4}\right| \qquad (2.20)$$

We have thus shown that the finite-difference scheme (2.10)–(2.11) does approximate differential problem (2.1)–(2.3) for its exact solution $u(x, t)$ with the limited second derivatives with respect to t and the limited fourth derivatives with respect to x. The approximations with respect to τ and h are of the first and second orders, respectively. It is usually said that the differencing scheme (2.10)–(2.11) approximates the problem (2.1)–(2.3) for the solution of the latter within $O(\tau + h^2)$.*

In the case under consideration, the boundary conditions and the right-hand side of the differential problem were approximated exactly and did not contribute to δf_h. However, if the boundary conditions or the right-hand side of the problem required an approximation, then this should have been included in the expression for δf_h.

---

*The expression $O(\alpha^n)$, for $n > 0$, denotes the value for which $\lim_{\alpha \to 0} [O(\alpha^n)/\alpha^n] = C$, where C is a nonzero constant.

2.3 STABILITY; CONVERGENCE THEOREM

In the previous sections, the problem of finding the exact solution $u(x, t)$ of the differential equation $Lu = f$ was reduced to compiling a table u_i^n of numerical values of the solution at the grid points. To calculate the table, the finite-difference scheme $L_h U_h = f_h$ was constructed to approximate the initial differential problem for the solution $u(x, t)$. The definition of the approximation was also given.

It should next be noted that even though a finite-difference scheme approximates the initial differential problem within its exact solution, we cannot state that the solution of the difference problem will converge to the exact solution of the differential problem as the grid steps become smaller.

Before demonstrating the validity of the above statement, it is necessary to define the term convergence. We shall say [2] that the solution U_h of the difference problem $L_h U_h = f_h$ converges to the exact solution $u(x, t)$ of the differential problem $Lu = f$ as the grid steps become smaller if, with vanishing τ and h, the difference $|U_h - u_h|$ also vanishes. If, in addition, the inequality

$$|U_h - u_h| \leqslant c_3 \tau^{k_3} + c_4 h^{k_4} \tag{2.21}$$

is fulfilled, where c_3, k_3 and c_4, k_4 are some positive constants independent of τ and h, then it is convergent of order $O(\tau^{k_3} + h^{k_4})$; in other words, the difference scheme is of the k_3th order of accuracy with respect to τ, and the k_4th order of accuracy with respect to h.

It should be noted that the convergence property is not the most important requirement for a difference scheme. If the difference scheme converges with some small enough τ and h, then the differential problem may be solved within any prescribed accuracy with the aid of the scheme.

We shall not demonstrate that any difference scheme is not necessarily convergent. As an illustration, let us turn back to the problem (2.1)-(2.3). We shall consider more carefully the difference scheme (2.10)-(2.11) which, written in a different form, becomes

$$U_i^{n+1} = (1 - 2\zeta)U_i^n + \zeta(U_{i+1}^n + U_{i-1}^n) + \tau Q_i^n$$
$$U_i^0 = T_i \quad U_0^n = T_0^n \quad U_J^n = T_1^n \tag{2.22}$$

where $\zeta = \kappa \tau / h^2$

BASIC CONCEPTS OF THE FINITE-DIFFERENCE METHOD

With the solutions U_i^n at $t^n = n\tau$ known, the values of U_i^{n+1} at a subsequent time $t^{n+1} = t^n + \tau$ may be calculated from the above formula. In other words, with the prescribed values of U_i^0 at initial time $t = 0$, the values of U_i^1, U_i^2, U_i^3, ... may be successively found with formula (2.22).

In an ideal case, the numerical solution U_i^n coincides with the exact solution $u(x, t)$ of the initial differential problem, or rather with a projection of the solution on the points of the grid, i.e., with the function $u_i^n = u(x_i, t^n)$. This situation is impossible, since the difference problem is always different from the original differential problem. Besides, the computations always involve truncation errors, and the solutions always differ from each other by a certain quantity

$$U_i^n = u_i^n + \delta u_i^n \tag{2.23}$$

Substitution into Eq. (2.22) gives

$$\frac{\delta u_i^{n+1} - \delta u_i^n}{\tau} - \frac{\kappa}{h^2}(\delta u_{i+1}^n - 2\delta u_i^n + \delta u_{i-1}^n) = Q_i^n$$

$$- \left[\frac{u_i^{n+1} - u_i^n}{\tau} - \frac{\kappa}{h^2}(u_{i+1}^n - 2u_i^n + u_{i-1}^n)\right] \tag{2.24}$$

Since the difference scheme (2.10) was earlier found to approximate the initial problem (2.1)-(2.3) with the order $O(\tau + h^2)$, the expression (2.24) may be written as

$$\frac{\delta u_i^{n+1} - \delta u_i^n}{\tau} = \frac{\kappa}{h^2}(\delta u_{i+1}^n - 2\delta u_i^n + \delta u_{i-1}^n) + O(\tau + h^2)$$

or (2.25)

$$\delta u_i^{n+1} = (1 - 2\zeta)\delta u_i^n + \zeta(\delta u_{i+1}^n + \delta u_{i-1}^n) + \tau O(\tau + h^2)$$

Relation (2.25) describes the evolution of the error in the transition from one time level to another. For a numerical scheme to be stable, the error made at one step of the computation should not be increased by subsequent computations.

First, consider the case in which $\zeta = \kappa\tau/h^2 > 0$ satisfies the inequality

$$0 \leqslant \zeta \leqslant 0.5 \qquad \tau \leqslant \frac{h^2}{2\kappa} \tag{2.26}$$

In this case, the coefficients of δu_i^n, δu_{i+1}^n, and δu_{i-1}^n are positive, and we may write

$$|\delta u_i^{n+1}| \leq (1 - 2\zeta)|\delta u_i^n| + \zeta(|\delta u_{i-1}^n| + |\delta u_{i+1}^n|) + \tau O(\tau + h^2)$$
$$\leq \max(1|\delta u_i^n|, |\delta u_{i+1}^n|, |\delta u_{i-1}^n|) + \tau O(\tau + h^2)$$

Since the above inequality holds for all indices i, the relation

$$\max_i |\delta u_i^{n+1}| \leq \max_i |\delta u_i^n| + \tau O(\tau + h^2)$$

holds; i.e., the maximum deviation δu for one step τ increases by not more than $\tau O(\tau + h^2)$. Accordingly, for N steps,

$$\max_i |\delta u_i^N| \leq \max_i |\delta u_i^0| + N\tau O(\tau + h^2)$$

In analyzing various practical problems, it is necessary to know their solutions over a finite time interval. Suppose we seek the solution of problem (2.1)–(2.3) for a time interval $0 \leq t \leq T' \leq N\tau$. With the value of T' fixed, and letting τ and h go to zero, it follows that N tends to infinity. Since the initial conditions of the original problem are exactly approximated $[\delta u_i^0 = u(x_i, 0) - u_i^0 = 0]$, then in the limit

$$\max_i |\delta u_i^N| = O(\tau + h^2)$$

Thus, it is proved that if, with vanishing τ and h, conditions (2.26) are fulfilled, the solution of the difference problem (2.10)–(2.11) converges to the solution of the original differential problem (2.1)–(2.3).

Now consider the case in which condition (2.26) breaks down. For example, let $\zeta > \frac{1}{2}$. Now it will be shown that for any $\zeta > \frac{1}{2}$ the difference scheme (2.10)–(2.11) is unstable, and the error made at some step of the computation will increase infinitely in subsequent computations.

If $\zeta > \frac{1}{2}$, it may easily be inferred that, as n increases, the sign of δu_i^n at x_i should alternate. The error δu_i^n is some complex function of the index i. But since the sign of the error varies from one mesh point to another, it may be expressed at any x_i as the sum with one addend of the form $(-1)^i \epsilon$. We shall trace the evolution of this error component alone.

Without limiting the generality, the error δu_i^0 may be assumed to be made at the initial time $t = 0$ as the approximation of the initial conditions. In view of the above, it may be assumed that

BASIC CONCEPTS OF THE FINITE-DIFFERENCE METHOD

$$\delta u_i^0 = (-1)^i \epsilon \tag{2.27}$$

Now let us return to Eq. (2.25) describing the error evolution from one time level to another. The contribution of the term $\tau O(\tau + h^2)$ in subsequent steps will be neglected (this, of course, is an idealization). Then, from Eq. (2.25), we find

$$\delta u_i^1 = (1 - 2\zeta)\delta u_i^0 + \zeta(\delta u_{i+1}^0 + \delta u_{i-1}^0)$$
$$= (1 - 2\zeta)(-1)^i \epsilon + \zeta(-1)^{i+1} \epsilon + \zeta(-1)^{i-1} \epsilon = (1 - 4\zeta)(-1)^i \epsilon$$

In the same way, we get

$$\delta u_i^2 = (1 - 4\zeta)^2 (-1)^i \epsilon$$
$$\delta u_i^3 = (1 - 4\zeta)^3 (-1)^i \epsilon$$
$$\cdots \cdots \cdots \cdots$$
$$\delta u_i^n = (1 - 4\zeta)^n (-1)^i \epsilon$$

Thus, the evolution of the error component will be determined by $(1 - 4\zeta)$. If

$$|1 - 4\zeta| \leqslant 1 \tag{2.28}$$

the errors are damping. In the opposite case, they increase exponentially. As the result, the solution U_i^n, containing the error, very quickly becomes meaningless, turning to a random succession of very large numbers. Decreasing both τ and h does not improve the situation but rather worsens it. Indeed, consider a finite time interval $T' = N\tau$. If condition (2.28) breaks down, then at $\tau, h \to 0$ and

$$|\delta u_i^N| = \left|1 - \frac{4\tau\kappa}{h^2}\right|^{T'/\tau} \epsilon \to \infty$$

Decreasing ϵ may only somewhat reduce the problem.

The above effect is called *instability*. It is quite clear that conditions (2.26) and (2.28) are equivalent. Thus, instability and the absence of convergence have one and the same cause.

In summary, the above analysis demonstrates that, as the condition

$$\tau \leqslant \frac{h^2}{2\kappa} \tag{2.29}$$

is satisfied, the solution of the difference problem (2.10)-(2.11) converges to the exact solution of the differential problem (2.1)-(2.3) as τ and h vanish. Otherwise, the difference scheme (2.10)-(2.11) gives unstable computations, and convergence may not be achieved.

One important fact should be emphasized: The difference scheme (2.10)-(2.11) approximates the original differential problem (2.1)-(2.3), irrespective of whether condition (2.20) is fulfilled. It appears, however, that a single approximation is not sufficient for the convergence of U_i^n and $u(x, t)$. Condition (2.29) is an additional condition that ensures convergence.

Thus we have come to the concept of the stability of a finite-difference scheme. The difference scheme $L_h U_h = f_h$ written for the approximate calculation of the differential problem $Lu = f$ is said to be *stable* if at any sufficiently small τ and h it has a unique solution U_h which is of the same order as the right-hand side f_h, that is, $U_h \approx f_h$. In other words, the difference scheme $L_h U_h = f_h$ is stable if the difference operator L_h is taken such that a small disturbance of the right-hand side f_h gives rise to a disturbance of the solution U_h of the same order.

This definition of the stability of a finite-difference scheme, presented in [1, 2], is closely connected with the concept of the correctness of continuous-argument problems. Stability may be said to establish a continuous dependence of the solution on the input data for discontinuous-argument problems as well.

It should be mentioned that, according to the above definition of stability, fulfillment of the stability conditions depends only on the properties of a finite-difference scheme and is quite independent of the original differential problem. Thus, stability is an intrinsic property of a difference problem.

The above convergence analysis procedure is typical for a wide range of problems and may be formulated as follows. Let some initial differential problem $Lu = f$ be replaced by the difference problem $L_h U_h = f_h$. Since the following discussion is a general one, no specific difference problem and difference scheme will be given. Here, L is simply a differential operator, and L_h is a linear difference operator. The functions f and f_h include not only the right-hand sides but also the boundary and initial conditions.

To show whether the solution U_h of the difference problem converges to the solution u_h of the original problem, it is assumed that

$$U_h = u_h + \delta u_h$$

Substitution of this expression into the difference problem gives the equation for δu_h:

$$L_h(\delta u_h) = f_h - L_h u_h \tag{2.30}$$

Convergence, or a vanishing δu_h, is achieved if, first, by an appropriate choice of the parameters τ and h, the right-hand side of (2.30) can be made infinitesimal, and, second, the infinitesimality is retained in the solution of Eq. (2.30) with respect to the grid function δu_h.

The first of the above conditions ($|f_h - L_h u_h| \to 0$ as $\tau, h \to 0$) is the familiar approximation condition that characterizes the relation between the difference and differential problems and establishes the similarity of the problems.

The second condition is the stability condition, which states that the difference operator L_h should have such a value that at any τ and h the solution of problem (2.30) is of the same order as the right-hand side:

$$\delta u_h \sim f_h - L_h u_h \tag{2.31}$$

Whether or not the above condition is satisfied depends only on the properties of the difference problem. No difference operator possesses a stability property. As was demonstrated above for the operator L_h from problem (2.10)-(2.11), condition (2.31) is fulfilled if κ, τ, and h satisfy relation (2.29).

In general, then, for any linear difference problem the following statement (the convergence theorem) holds: If a difference boundary-value problem approximates a differential problem within its exact solution and is stable, the solution of the difference problem converges to the exact solution of the differential problem as the step sizes decrease, and the order of the accuracy of the difference scheme is the same as the order of approximation.

This is one of the fundamental theorems of computational mathematics. It was formulated in the middle fifties almost simultaneously by Ryabenky and Filippov [1] and Lax and Richtmyer [3], who approached the subject from different points of view. The proof of the theorem will not be presented, as it is beyond the scope of this book.

The convergence theorem implies that, in selecting a convergent difference scheme for the numerical solution of a differential problem, one should choose, from the many available numerical methods, that approximation for which the difference scheme is stable.

2.4 STABILITY ANALYSIS OF FINITE-DIFFERENCE COMPUTATIONS BY THE FOURIER TRANSFORMATION

Whereas it is not difficult to study the order of approximation even for complicated problems, testing the stability of finite-difference procedures is extremely difficult. In very simple cases, the problem is solvable by elementary means. There is extensive literature on the stability analysis of finite-difference schemes. It is practically impossible to test the stability of difference algorithms for nonlinear problems, and therefore the stability of finite-difference procedures may be checked only for linear problems.

Let us consider one of the most common methods of stability analysis, based on the Fourier transformation and suggested in [4]. Generally speaking, this method is applicable only to linear problems with constant coefficients at the boundary conditions, for which periodicity conditions may be substituted. In particular cases, this method may be extended to problems of a more general nature.

The stability analysis of finite-difference computations by the Fourier transformation will be illustrated with a simple Cauchy problem for the heat-conduction equation:

$$\frac{\partial u}{\partial t} = \kappa \frac{\partial^2 u}{\partial x^2} \qquad -\infty < x < \infty \qquad t > 0 \tag{2.32}$$

$$u(x, 0) = \psi(x) \qquad \kappa = \text{const} > 0$$

For the approximate calculation of the solution of (2.32), a uniform grid is introduced. The grid points have coordinates $(x_i = ih, \ t^n = n\tau)$, where h and τ are the discrete space and time grid steps. The finite-difference scheme is written as

$$\frac{U_k^{n+1} - U_k^n}{\tau} = \frac{\kappa}{h^2}(U_{k+1}^n - 2U_k^n + U_{k-1}^n) \tag{2.33}$$

$$U_k^0 = \psi_k \qquad k = 0, \pm 1, \pm 2, \ldots \qquad n = 0, 1, 2, \ldots$$

The functions $U_k^1, U_k^2, \ldots$ may successively be calculated with the formula

$$U_k^{n+1} = (1 - 2\zeta)U_k^n + \zeta(U_{k+1}^n + U_{k-1}^n) \qquad \zeta = \frac{\kappa\tau}{h^2} \tag{2.34}$$

To emphasize the general validity of the present problem, expression (2.34) will be written in operator form as

$$U_k^{n+1} = SU_k^n \tag{2.35}$$

where S is a linear operator, dependent on the grid parameters τ and h, which transforms the function U_k^n, defined on the layer n, to the function U_k^{n+1}, defined on the $(n+1)$th layer. Here the term "layer" is used to denote the unity of the grid points lying on the straight line $t = \text{const}$. Operator S, which will be referred to as a *transition* operator, may be written as

$$SU_k^n = (1 - 2\varsigma)U_k^n + \varsigma(U_{k+1}^n + U_{k-1}^n)$$

or, in an abbreviated form, as

$$SU_k^n = \sum_{\alpha=-1}^{1} a_\alpha U_{k+\alpha}^n \tag{2.36}$$

where the a_α's are the coefficients dependent on τ and h.

It should be noted that a polynomial form for a transition operator is typical for most linear problems. The number of terms in formula (2.36) may vary, depending on the finite-difference scheme adopted.

Since the operator S is linear and independent of n, for a certain fixed, large time $T = N\tau$, we may write, according to (2.35),

$$U_k^N = SU_k^{N-1} = S^2 U_k^{N-2} = \cdots = S^N U_k^0 \tag{2.37}$$

In accordance with the above definition of stability, the difference procedure (2.35) is stable if at any U_k^0 the following approximation holds:

$$|U_k^n| \leqslant C|U_k^0| \tag{2.38}$$

where $C > 0$ is a constant independent of τ and h. In the case under consideration, stability condition (2.38) is equivalent to a uniform limitation on the degrees of the transition operator

$$|S^n| \leqslant C \quad n = 1, 2, 3, \ldots \tag{2.39}$$

The validity of condition (2.39) may readily be checked with the aid of the Fourier transformation applied to a finite-difference scheme, if the coefficients of the scheme are independent of x and t.

It should be remembered that the Fourier transform of a function $v(x)$ is the function $F(\omega)$ of the real variable $\omega \in (-\infty, \infty)$ defined by the identity [7]

$$F(\omega) = \frac{1}{\sqrt{2\pi}} \int_{-\infty}^{\infty} e^{-i\omega x} v(x) \, dx \qquad (2.40)$$

The Fourier transform of the function $v(x)$ is sometimes denoted by $F_\omega[v(x)]$.

Let us calculate the Fourier transforms of particular functions. The function $v(x + h)$ will first be considered. From the definition,

$$F_\omega[v(x + h)] = \frac{1}{\sqrt{2\pi}} \int_{-\infty}^{\infty} e^{-i\omega x} v(x + h) \, dx$$

$$= \frac{1}{\sqrt{2\pi}} \int_{-\infty}^{\infty} e^{-i\omega(y-h)} v(y) \, dy = e^{i\omega h} F_\omega[v(x)] \qquad (2.41)$$

According to (2.41), the Fourier-transform expression may easily be obtained for different operators also:

$$F_\omega[v(x - h)] = e^{-i\omega h} F_\omega[v(x)] \qquad (2.42)$$

$$F_\omega\left[\frac{v(x+h) - v(x)}{h}\right] = \frac{e^{i\omega h} - 1}{h} F_\omega[v(x)] \qquad (2.43)$$

$$F_\omega\left[\frac{v(x) - v(x-h)}{h}\right] = \frac{1 - e^{-i\omega h}}{h} F_\omega[v(x)] \qquad (2.44)$$

$$F_\omega\left\{\frac{1}{h^2}[v(x+h) - 2v(x) + v(x-h)]\right\} = -\frac{4\sin^2(\omega h/2)}{h^2} F_\omega[v(x)] \qquad (2.45)$$

Now let us return to the difference scheme (2.35) and apply the Fourier transformation. Generally speaking, the application of the Fourier transformation to a grid function requires the assumption of smooth

BASIC CONCEPTS OF THE FINITE-DIFFERENCE METHOD 25

extension of the function to the whole straight line $-\infty < x < \infty$. In most practical cases, this assumption is quite realistic. Canceling the common factor $\exp(i\omega kh)$ results in

$$V_{(\omega)}^{n+1} = \lambda(\omega) V_{(\omega)}^n \qquad (2.46)$$

which will be referred to as the *Fourier transform* of the difference scheme (2.35). Here $V(\omega)$ and $\lambda(\omega)$ are used to denote the Fourier transforms of the function U and the operator S, respectively. Compared with (2.35), the relation (2.46) has an essentially simpler structure. It contains no space difference operators, and a transition coefficient $\lambda(\omega)$ is substituted for the transition operator S.

Since, from mathematical analysis, the Fourier transform is known to retain its metric properties of operators, the stability condition (2.39) of a difference scheme may now be formulated as a limitation on the transition coefficient

$$|\lambda^n(\omega)| \leq c \quad \text{at } \tau, h \to 0 \qquad (2.47)$$

Inequality (2.47) should certainly be satisfied at all the values of ω. Transition coefficient $\lambda(\omega)$ depends on parameters τ and h, and since at a fixed t, $n = t/\tau$ tends to infinity as τ decreases, condition (2.47) is equivalent to the limitation

$$\max_{\omega} |\lambda(\omega)| \leq 1 + c\tau \quad \text{at } \tau, h \to 0 \qquad (2.48)$$

Indeed, if $|\lambda| = 1 + c\tau$, then

$$|\lambda^n| \sim (1 + c\tau)^{t/\tau} \sim e^{ct}$$

Although constant, the above value cannot be too large. Therefore, instead of (2.48), the stronger limitation

$$\max_{\omega} |\lambda(\omega)| \leq 1 \quad \text{at } \tau, h \to 0 \qquad (2.49)$$

is sometimes considered.

The transition coefficient for the difference scheme (2.33) may be obtained by replacing U_{k+2}^n with a complex number $\exp(i\omega\alpha h)$ as

$$\lambda(\omega) = 1 - \frac{4\kappa\tau}{h^2}\sin^2\frac{\omega h}{2}$$

Condition (2.49) evidently is satisfied, and difference scheme (2.33) is therefore stable, if

$$\tau \leqslant \frac{h^2}{2\kappa} \tag{2.50}$$

It should be noted that the transition factor $\lambda(\omega)$ may be found by direct substitution of the expression $\lambda^n \exp(i\omega kh)$ for $U_k{}^n$ in the difference equation. Such an approach is very useful in a stability analysis of implicit difference schemes which cannot easily be expressed in the form (2.35).

Let us consider, for example, a finite-difference scheme of the form

$$\frac{U_k{}^{n+1} - U_k{}^n}{\tau} = \frac{\kappa}{h^2}(U_{k+1}^{n+1} - 2U_k{}^{n+1} + U_{k-1}^{n+1}) \quad U_k{}^0 = \psi_k \tag{2.51}$$

This difference scheme approximates problem (2.32) and is defined for the same grid as scheme (2.33). For stability analysis, the expression $\lambda^n \exp(i\omega kh)$ is substituted for $U_k{}^n$. Appropriate cancellations result in

$$\frac{\lambda - 1}{\tau} = \frac{\kappa\lambda}{h^2}(e^{i\omega h} - 2 + e^{-i\omega h})$$

Hence,

$$\lambda(\omega) = \left[1 + 4\zeta\sin^2\left(\omega\frac{h}{2}\right)\right]^{-1} \quad \zeta = \frac{\kappa\tau}{h^2}$$

The condition $|\lambda(\omega)| \leqslant 1$ is fulfilled at any τ and h, and scheme (2.51) is therefore absolutely stable.

Stability analysis by the Fourier transformation may easily be extended to cases of two and more dimensions. As an illustration, consider the problem

$$\frac{\partial u}{\partial t} = \kappa\left(\frac{\partial^2 u}{\partial x^2} + \frac{\partial^2 u}{\partial y^2}\right) \quad -\infty < x, y < \infty \quad t > 0 \tag{2.52}$$

$$u(x, y, 0) = \psi(x, y) \quad\quad \kappa = \text{const} > 0$$

BASIC CONCEPTS OF THE FINITE-DIFFERENCE METHOD

Let us introduce a grid whose points have coordinates $(x_k = kh, y_l = lh, t^n = n\tau)$ and write the approximating difference scheme as

$$\frac{U_{k,l}^{n+1} - U_{k,l}^n}{\tau} = \frac{\kappa}{h^2}(U_{k+1,l}^n + U_{k-1,l}^n + U_{k,l+1}^n + U_{k,l-1}^n - 4U_{k,l}^n)$$

$$U_{k,l}^0 = \psi_{k,l}$$
(2.53)

Here the notation $U_{k,l}^n = U(x_k, y_l, t^n)$ is used. Stability analysis involves the substitution of the expression $\lambda^n(\omega_1, \omega_2) \exp(i\omega_1 kh + i\omega_2 lh)$ for $U_{k,l}^n$ in the difference equation. Cancellations and identical transformations give

$$\lambda(\omega_1, \omega_2) = 1 - 4\zeta \sin^2 \frac{\omega_1 h}{2} - 4\zeta \sin^2 \frac{\omega_2 h}{2} \qquad \zeta = \frac{h^2}{4\kappa\tau}$$

Difference scheme (2.53) is stable if $|\lambda(\omega_1, \omega_2)| \leq 1$ at any real ω_1 and ω_2 that are possible at $\tau \leq h^2/4\kappa$.

All considerations concerning the application of the Fourier transformation to the stability analysis of finite-difference schemes that approximate one differential equation are valid for systems of differential equations. In such cases, the transition operator S of the difference scheme is represented by a square $p \times p$ matrix, where p is the number of equations in the initial system. Accordingly, the application of the Fourier transformation to the difference scheme results in the substitution of a transition matrix $\hat{S}(\omega)$ for the transition factor $\lambda(\omega)$, the matrix being written in terms of real or complex numbers.

The stability condition is formulated in the form of limitations on the transition matrix $\hat{S}(\omega)$, and it is equivalent to a limitation of the degrees of the matrix. Since the matrix value is characterized by its eigenvalues, the stability condition is

$$\max_\omega \max_m |\lambda_m[\hat{S}(\omega)]|^n \leq c \qquad 1 \leq m \leq p \tag{2.54}$$

where $\lambda_m[\hat{S}(\omega)]$ is the mth eigenvalue of the matrix $\hat{S}(\omega)$. Since the matrix $\hat{S}(\omega)$ is complex,

$$\lambda_m[\hat{S}(\omega)] = \sqrt{\overline{\lambda_m(\hat{S}^*S)}} \tag{2.55}$$

where $\hat{S}^*$ is the complex conjugate of $\hat{S}$ in a complex way.

The eigenvalues of the transition matrix $\hat{S}(\omega)$ could also be found by the substitution of expressions of the form

$$u_1{}^0 \lambda^n \exp(i\omega kh), \; u_2{}^0 \lambda^n \exp(i\omega kh), \ldots, u_p{}^0 \lambda^n \exp(i\omega kh)$$

for the grid functions $U_{1,k}^n, U_{2,k}^n, \ldots, U_{p,k}^n$ in the difference equations. The substitution results in a system of p linear equations for $u_1{}^0, u_2{}^0, \ldots, u_p{}^0$. The eigenvalues are determined from the assumption of a zero determinant composed of the coefficients of the system.

It should be noted that the stability analysis of many-layered finite-difference procedures is in many cases facilitated by the reduction of the finite-difference schemes to systems of two-layered difference equations. Consider, for example, the well-known finite-difference scheme suggested by DuFort and Frankel [5] for the solution of heat-conduction equation (2.32):

$$\frac{U_k^{n+1} - U_k^{n-1}}{2\tau} = \frac{\kappa}{h^2}(U_{k+1}^n - U_k^{n+1} - U_k^{n-1} + U_{k-1}^n) \qquad (2.56)$$

We shall return to the scheme more than once, but at present we are interested in the stability of the scheme. It will be demonstrated that difference equation (2.56) is always stable. For this purpose, the equivalent two-layered system is written

$$\begin{aligned} U_k^{n+1} &= V_k^n + \frac{2\kappa\tau}{h^2}(U_{k+1}^n - U_k^{n+1} - V_k^n + U_{k-1}^n) \\ V_k^{n+1} &= U_k^n \end{aligned} \qquad (2.57)$$

The eigenvalues of the transition matrix for difference scheme (2.57) are found by the substitutions

$$U_k^n = u_0 \lambda^n \exp(i\omega kh) \qquad V_k^n = v_0 \lambda^n \exp(i\omega kh)$$

Then, a simple transformation results in the following system:

$$\begin{aligned} [2\varsigma \cos \omega h - \lambda(1+\varsigma)]u_0 + (1-\varsigma)v_0 &= 0 \\ u_0 - \lambda v_0 &= 0 \end{aligned} \qquad (2.58)$$

BASIC CONCEPTS OF THE FINITE-DIFFERENCE METHOD

For the homogeneous system (2.58) to have a solution different from zero, the determinant composed of the coefficients of the system should be zero. Thus we have

$$\lambda^2(1+\zeta) - 2\lambda\zeta \cos \omega h - \zeta + 1 = 0$$

Solution of this equation gives the eigenvalues of the transition matrix in the form

$$\lambda_{1,2} = \frac{\zeta \cos \omega h \pm \sqrt{1 - \zeta^2 \sin^2 \omega h}}{1 + \zeta}$$

It may easily be verified that at any ω, τ, and h the moduli of eigenvalues λ_1 and λ_2 will be less than unity, and the difference scheme (2.56) is therefore always stable.

It should be noted here that the Fourier transformation reduces the stability analysis to the purely algebraic problem of finding the conditions for the limitation of the degrees of the transition matrix $\hat{S}(\omega)$. The simplest of the conditions is

$$|\lambda_m[\hat{S}(\omega)]| \leq 1 + c\tau \quad \text{at } \tau, h \to 0 \qquad (2.59)$$

where $\lambda_m[\hat{S}(\omega)]$ is the mth eigenvalue of the transition matrix $\hat{S}(\omega)$, and c is a constant independent of ω, τ, and h. The condition discussed above is known in the literature as the *Neumann condition*.

The Neumann condition is the necessary spectral stability condition. It is called "spectral" because the whole line, passed by point $\lambda(\omega)$ at a complex plane as ω passes the whole real axis, consists of eigenvalues and is the spectrum of the transition operator.

It should be noted that (2.59) is a necessary and sufficient condition only for the stability of a finite-difference scheme once the matrix $\hat{S}(\omega)$ is a normal matrix, that is, $\hat{S}^*\hat{S} = \hat{S}\hat{S}^*$. In particular, the Neumann condition is necessary and sufficient for two-layered finite-difference schemes with one dependent variable, as in this case the transition matrix is of the first order and all such matrices are permutable. In other cases, the inequality (2.59) is only necessary for stability.

Much of the literature is concerned with the necessary and sufficient conditions for limitations on transition-matrix degrees, and some efforts in this regard have been successful. However, the conditions given in the literature, while useful for theoretical considerations, cannot, as a rule, easily be used directly for a particular scheme.

In the case of variable coefficients and boundary conditions that are essentially nonperiodic, direct application of the Fourier transformation is impossible, and the problem cannot be reduced to an algebraic one. In that situation, the principle of "frozen" coefficients may sometimes be useful. The essence of this principle consists of fixing the variables in the coefficients of the scheme, subject to subsequent stability analysis of the resultant scheme, with constant coefficients. If the scheme with "frozen" coefficients appears to be stable at any x_0, then the scheme with variable coefficients is also considered stable.

In this case, the stability condition is more severe than appropriate schemes with constant coefficients. For example, it was found above that for difference scheme (2.33), which approximates problem (2.32), the Neumann condition is satisfied if the ratio of the grid steps meets the condition $\tau \leqslant h^2/2\kappa$. Because of the principle of "frozen" coefficients for stability of the problem with variable coefficients [say, $\kappa(x)$], this condition should be satisfied at any x. Thus, this condition should be replaced with the condition

$$\tau \leqslant \frac{h^2}{2 \max_x \kappa(x)} \tag{2.60}$$

Note, however, that it has been demonstrated that the stability of a scheme with "frozen" coefficients is generally neither a necessary nor a sufficient stability condition for a scheme with variable coefficients.

2.5 STABILITY ANALYSIS OF FINITE-DIFFERENCE COMPUTATIONS BY THE MAXIMUM PRINCIPLE

The maximum principle may very often be useful for the stability analysis of finite-difference procedures. Unlike the analysis described in the previous section, the procedure considered here allows stability analysis not only for Cauchy problems but also for problems with boundary conditions. The method is superior, since it allows the stability analysis of difference procedures for problems with variable coefficients. It is quite popular because of its relative simplicity.

The use of the maximum principle for stability analysis will be illustrated with two examples. These examples concern the stability of

BASIC CONCEPTS OF THE FINITE-DIFFERENCE METHOD

explicit and implicit finite-difference schemes for the heat-conduction problem with variable coefficients [2]:

$$\frac{\partial u}{\partial t} - \kappa(x, t)\frac{\partial^2 u}{\partial x^2} = \varphi(x, t) \quad 0 \leqslant x \leqslant 1 \quad t > 0$$

$$u(x, 0) = \psi(x) \quad \kappa(x, t) > 0 \qquad (2.61)$$

$$u(0, t) = \psi_1(t) \quad u(1, t) = \psi_2(t)$$

The explicit scheme

$$\frac{U_k^{n+1} - U_k^n}{\tau} - \frac{\kappa_k^n}{h^2}(U_{k+1}^n - 2U_k^n + U_{k-1}^n) = \varphi_k^n$$

$$U_k^0 = \psi_k \quad U_0^n = \psi_1^n \quad U_K^n = \psi_2^n \qquad (2.62)$$

$$k = 1, 2, 3, \ldots, K-1 \quad n = 0, 1, 2, 3, \ldots$$

will be considered first. The maximum principle for this scheme will now be determined. To this end, the difference equation that is the basis of the scheme is rewritten as

$$U_k^{n+1} = (1 - 2\zeta\kappa_k^n)U_k^n + \zeta\kappa_k^n(U_{k+1}^n + U_{k-1}^n) + \tau\varphi_k^n \quad \zeta = \frac{\tau}{h^2}$$

If

$$\tau \leqslant \frac{h^2}{2 \max_{k,n} \kappa_k^n} \qquad (2.63)$$

then the factor of U_k^n is nonnegative. The inequality may therefore be written

$$|U_k^{n+1}| \leqslant (1 - 2\zeta\kappa_k^n) \max_k |U_k^n| + 2\zeta\kappa_k^n \max_k |U_k^n| + \tau \max_{k,n} |\varphi_k^n|$$

$$= \max_k |U_k^n| + \tau \max_{k,n} |\varphi_k^n|$$

With $U_0^{n+1} = \psi_1^{n+1}$ and $U_K^{n+1} = \psi_2^{n+1}$, the above inequality is the maximum principle for the explicit difference scheme:

$$\max_k |U_k^{n+1}| \leqslant \max\left(\max_n |\psi_1^n|, \max_n |\psi_2^n|, \max_k |U_k^n| + \tau \max_{k,n} |\varphi_k^n|\right)$$

$$(2.64)$$

Indeed, as $\varphi_k^n \equiv 0$ and $\psi_1, \psi_2 = $ const, inequality (2.64) implies that $\max_k |U_k^n|$ does not increase with n. This property of the difference scheme is called the *maximum principle*, but the term is generally used to denote the whole inequality (2.64).

It will be demonstrated that if a difference scheme satisfies the maximum principle (2.64), it is stable. Recall the definition of stability: Difference scheme $L_h U_h = f_h$ is stable if at any sufficiently small h and any f_h it has a unique solution and the following inequality holds:

$$|U_h| \leqslant c|f_h| \tag{2.65}$$

where $c > 0$ is independent of τ and h. In the present case, $|U_h| = |U_k^n|$, and $|f_h| = |\psi_1^n|, |\psi_2^n|, |\psi_k|,$ or $|\varphi_k^n|$.

To study the stability, we express the solution U_k^n as the sum of two terms, V_k^n and W_k^n. The term V_k^n is defined as the solution of the difference problem (2.62) at $\varphi_k^n = 0$, and the term W_k^n as the solution of the same problem with zero initial and boundary conditions, i.e., at $\psi_k = \psi_1^n = \psi_2^n = 0$.

Since the maximum principle (2.64) is valid for the difference scheme (2.62) for any $\varphi_k^n, \psi_k, \psi_1^n,$ and ψ_2^n, it may easily be written as

$$\max_k |V_k^{n+1}| \leqslant \max(\max_n |\psi_1^n|, \max_n |\psi_2^n|, \max_k |V_k^n|)$$

$$\max_k |V_k^n| \leqslant \max(\max_n |\psi_1^n|, \max_n |\psi_2^n|, \max_k |V_k^{n-1}|)$$

$$\cdots\cdots\cdots\cdots\cdots\cdots\cdots\cdots\cdots\cdots\cdots$$

$$\max_k |V_k^1| \leqslant \max(\max_n |\psi_1^n|, \max_n |\psi_2^n|, \max_k |\psi_k|)$$

That is,

$$\max_k |V_k^{n+1}| \leqslant \max(\max_n |\psi_1^n|, \max_n |\psi_2^n|, \max_k |\psi_k|)$$

To solve for W_k^{n+1} using (2.64), we have

$$\max_k |W_k^{n+1}| \leqslant \max_k |W_k^n| + \tau \max_{k,n} |\varphi_k^n| \leqslant \max_k |W_k^{n-1}| + 2\tau \max_{k,n} |\varphi_k^n|$$

$$\leqslant \cdots \leqslant \max_k |W_k^0| + (n+1)\tau \max_{k,n} |\varphi_k^n| \leqslant T \max_{k,n} |\varphi_k^n|$$

BASIC CONCEPTS OF THE FINITE-DIFFERENCE METHOD

Thus, for the solution of U_k^{n+1}, we have the approximation

$$\max_k |U_k^{n+1}| \leqslant \max_k |V_k^{n+1}| + \max_k |W_k^{n+1}|$$

$$\leqslant \max(\max_n |\psi_1^n|, \max_n |\psi_2^n|, \max_k |\psi_k|) + T \max_{k,n} |\varphi_k^n|$$

which is equivalent to approximation (2.65). Thus, the difference scheme (2.62) is really stable once condition (2.63) is satisfied.

Now we consider the implicit scheme

$$\frac{U_n^{k+1} - U_k^n}{\tau} - \frac{\kappa_k^n}{h^2}(U_{k+1}^{n+1} - 2U_k^{n+1} + U_{k-1}^{n+1}) = \varphi_k^n \quad (2.66)$$

$$U_0^{n+1} = \psi_1^{n+1} \qquad U_K^{n+1} = \psi_2^{n+1} \qquad U_k^0 = \psi_k$$

It may easily be demonstrated that U_k^{n+1} can be determined by solving the system

$$\varsigma\kappa_k^n U_{k-1}^{n+1} - (1 + 2\varsigma\kappa_k^n)U_k^{n+1} + \varsigma\kappa_k^n U_{k+1}^{n+1} = -U_k^n - \tau\varphi_k^n \quad (2.67)$$

$$U_0^{n+1} = \psi_1^{n+1} \qquad U_K^{n+1} = \psi_2^{n+1} \qquad U_k^0 = \psi_k$$

The determinant of the system is not zero, and it therefore has a single solution which may be found, say, by Gauss elimination.

To check the stability, one should demonstrate that inequality (2.65) is fulfilled. To this end, we shall prove that scheme (2.66) is consistent with the maximum principle (2.64). Then, the approximation of (2.65) can be found exactly as for an explicit scheme.

Consider U_k^{n+1}. Of all the values of U_k^{n+1} with the moduli of $\max_k |U_k^{n+1}|$, we choose the one whose index k has the smallest value, $k = k'$. If $k' = 0$ or $k' = K$, then (2.64) is obviously fulfilled. Let $k' \neq 0$ and $k' \neq K$. Then write the equation for this index:

$$\varsigma\kappa_{k'}^n(U_{k'-1}^{n+1} + U_{k'+1}^{n+1}) - (1 + 2\varsigma\kappa_{k'}^n)U_{k'}^{n+1} = -U_{k'}^n - \tau\varphi_{k'}^n$$

For definiteness, $U_{k'}^{n+1} > 0$ will be assumed. The left-hand side of the inequality may then be approximated as

$$\varsigma\kappa_{k'}^n[(U_{k'-1}^{n+1} - U_{k'}^{n+1}) + (U_{k'+1}^{n+1} - U_{k'}^{n+1})] - U_{k'}^{n+1} \leqslant -U_{k'}^{n+1}$$

But then

$$-U_{k'}{}^{n+1} \geqslant -U_{k'}{}^{n} - \tau\varphi_{k'}{}^{n}$$

Hence,

$$\max_{k} |U_k{}^{n+1}| = U_{k'}{}^{n+1} \leqslant |U_{k'}{}^{n} + \tau\varphi_{k'}{}^{n}| \leqslant \max_{k} |U_k{}^{n}| + \tau \max_{k,n} |\varphi_k{}^{n}|$$

Thus, inequality (2.64), which implies the validity of the maximum principle, is proved, meaning that the stability of scheme (2.66) is also proved.

There are other ways to study the stability of finite-difference procedures. The method of energy inequalities is a powerful instrument for stability analysis, as it allows one to account for variable coefficients and boundary conditions. In addition, the method may be used both for the stability analysis of a particular finite-difference procedure and to choose a suitable procedure.

The method involves the choice of a metric for solution such that, for each step, the solution increases by not more than $I + 0(\tau)$ times; i.e., the solution should be stable for the chosen metric. Additional considerations require that the chosen metric be equivalent to the energy norm. A detailed description of the method is beyond the scope of this book.

A very general theory of difference schemes based on energy inequalities was developed by Samarsky [6, 7]. Although we shall not discuss the energy method, it should be emphasized that this method does have a wide range of use in solving heat-transfer problems.

To conclude this section, it should be noted that, for nonlinear equations, no stability analysis procedures are available as yet for general classes of finite-difference schemes. However, experience with such problems shows that many of them may successfully be solved on existing computers. Here, there is a gap between the theory of difference procedures and its application to particular physical problems. Very often, success in solving a complicated problem results from a proper mix of physical intuition, empirical data, and the design of the numerical experiment.

2.6 CONSTRUCTION OF CALCULATION FORMULAS

At this point, the main requirements for finite-difference schemes should be quite clear, and we shall proceed to the construction of such schemes.

BASIC CONCEPTS OF THE FINITE-DIFFERENCE METHOD

The formulation of a difference problem begins with the choice of a differencing grid—a discrete set of points—in place of the continuous range of the independent variables of the original differential problem. In dealing with partial differential equations, there are a number of practical and theoretical problems for which solutions are not always known.

Among these problems, that of the optimal choice of grid parameters to characterize the mesh spacing is of primary importance. Certainly, for greater accuracy in numerical calculations, a small mesh spacing is preferable. However, accuracy requirements often conflict with other important factors, such as computer time and the possibility of realizing the program on a computer. Moreover, not every existing finite-difference scheme allows an independent choice of grid parameters. For example, for most explicit schemes, stability of the computation procedure is provided when $\tau \approx h^2$. A differencing grid based on stability conditions for a difference scheme should provide the required numerical accuracy with different computer times.

Having estimated the order of approximation for a finite-difference scheme, we may choose mesh sizes that would roughly assure the prescribed accuracy. However, it should be remembered that practical calculations are carried out with finite values of τ and h. In the solution of complicated problems, large discrete-space-variable steps are usually chosen ($\frac{1}{40} \leq h \leq \frac{1}{10}$). Therefore, in estimating the errors, the values of the derivatives involved in its main terms should also be taken into account. Since every physical problem has its own peculiarities, this is hardly possible before computation, when the solution is unknown.

The optimum values of the grid parameters are generally determined from numerical experimentation. First, the problem is solved by assuming large mesh spacing. The computation is then repeated for a grid with a small mesh size, and the results are compared. If the comparison shows a large difference, the computation is repeated for grids of smaller and smaller mesh sizes, until the required accuracy is achieved.

If the computations appear unstable during these numerical experiments, it is necessary to decrease the time step (by about an order), without changing the space steps.

The regularity of the grid is also important. A grid is called *regular* if it is uniform with respect to each of its parameters. In the majority of problems, the solution changes more quickly in some regions than in others. It is quite natural to choose a smaller mesh size for regions where the solution changes more quickly. However, the introduction of an irregular grid makes the algorithm more complicated. Therefore, such grids are usually introduced only when available regular grids do not provide the required accuracy.

In choosing a differencing grid, it is very important that the boundaries of the grid region coincide with the boundaries of the appropriate continuous region, since practical computations have shown that noncoincidence of the boundaries has a very appreciable effect on the accuracy of the resulting solution. Exact coincidence of the boundaries is usually achieved by a suitable choice of the mesh (square, rectangular, triangular, etc.) or by the introduction of a curvilinear grid, which often involves grid irregularity. When the boundaries do not coincide, the values prescribed for the function at the boundary mesh points are the values it has at the neighboring points of the real boundary.

Having chosen, we may now proceed to the construction of the finite-difference scheme. Some very simple means of construction for approximate difference schemes will be discussed.

The simplest method of constructing difference equations, that most often used in practice, is based on the substitution of linear combinations of the function values at the mesh points for the derivatives entering into the initial differential equation, with the aid of appropriate equations of numerical differentiation. The difference schemes described in previous sections were constructed by this method.

The formulas most frequently used for substituting difference relations for differential operators will now be presented, and the order of magnitude of the error will be given. For simplicity, the function $u(x, t)$ entering into an equation is assumed to depend only on two independent variables x and t.

The simplest differential operator, $\partial u/\partial x$, may be replaced with any of the following expressions (difference operators):

$$\frac{\partial u}{\partial x} \approx \frac{u(x + h, t) - u(x, t)}{h} \qquad (2.68)$$

$$\frac{\partial u}{\partial x} \approx \frac{u(x, t) - u(x - h, t)}{h} \qquad (2.69)$$

Expressions (2.68) and (2.69) are called the *forward* and *backward difference derivatives*, respectively. In what follows they will sometimes be denoted by $\delta_x^+ u$ and $\delta_x^- u$. The difference operators $\delta_x^+ u$ and $\delta_x^- u$ are determined at two points of the grid. In this case, it is usually said that the difference operators are determined from a two-point pattern.

BASIC CONCEPTS OF THE FINITE-DIFFERENCE METHOD

Naturally, any linear combination of expressions (2.68) and (2.69) will also approximate the derivative:

$$\frac{\partial u}{\partial x} \approx \sigma \delta_x^+ u + (1-\sigma)\delta_x^- u$$

where σ is any real number. Particularly at $\sigma = 0.5$, a central-difference derivative is obtained as

$$\delta_x u = 0.5(\delta_x^+ u + \delta_x^- u) = \frac{u(x+h, t) - u(x-h, t)}{2h} \quad (2.70)$$

Thus, to approximate the derivative $\partial u/\partial x$, an infinite set of difference relations may be written. If the function $u(x, t)$ is assumed to be smooth enough and to have a sufficient number of limited derivatives, then a Taylor-series expansion of the function about (x, t) readily gives the expressions for residual terms in formulas (2.68)–(2.70).

$$\delta_x^+ u = \frac{u(x+h, t) - u(x, t)}{h} = \frac{\partial u}{\partial x} + \frac{h}{2}\frac{\partial^2 u}{\partial x^2} + O(h^2)$$

$$\delta_x^- u = \frac{u(x, t) - u(x-h, t)}{h} = \frac{\partial u}{\partial x} - \frac{h}{2}\frac{\partial^2 u}{\partial x^2} + O(h^2)$$

$$\delta_x u = \frac{u(x+h, t) - u(x-h, t)}{2h} = \frac{\partial u}{\partial x} + \frac{h^2}{3}\frac{\partial^3 u}{\partial x^3} - O(h^4)$$

Thus, difference operators $\delta_x^- u$ and $\delta_x^+ u$ approximate the derivative $\partial u/\partial x$ with an order of $O(h)$, and operator $\delta_x u$ with an order $O(h^2)$.

To approximate boundary conditions, it may often be useful to substitute, for the derivative $\partial u/\partial x$, three-point difference operators of the form

$$-\frac{3u(x, t) - 4u(x+h, t) + u(x+2h, t)}{2h} = \frac{\partial u}{\partial x} - \frac{h^2}{3}\frac{\partial^3 u}{\partial x^3} + O(h^3) \quad (2.71)$$

or

$$\frac{3u(x, t) - 4u(x-h, t) + u(x-2h, t)}{2h} = \frac{\partial u}{\partial x} - \frac{h^2}{3}\frac{\partial^3 u}{\partial x^3} + O(h^3) \quad (2.72)$$

which are also of the second order of approximation.

The second-order partial derivative $\partial^2 u/\partial x^2$ is approximated in most cases by a three-point difference operator of the form

$$\delta_x^2 u = \frac{1}{h}\left[\frac{u(x+h,t)-u(x,t)}{h} - \frac{u(x,t)-u(x-h,t)}{h}\right]$$

$$= \frac{1}{h^2}[u(x+h,t) - 2u(x,t) + u(x-h,t)] = \frac{\partial^2 u}{\partial x^2} + \frac{h^2}{12}\frac{\partial^4 u}{\partial x^4} + O(h^3) \quad (2.73)$$

Now consider the approximation of the differential operator $\partial^2 u/\partial x^2$ for an irregular pattern (a nonuniform grid). For a three-point pattern $(x_i - h_1, x_i, x_i + h_2)$, where $h_1 \neq h_2$, a difference operator $\bar{\delta}_x^2 u$ with structure similar to that of the operator $\delta_x^2 u$ will be considered:

$$\bar{\delta}_x^2 u = \frac{2}{h_1 + h_2}\left[\frac{u(x+h_2,t) - u(x,t)}{h_2} - \frac{u(x,t) - u(x-h_1)}{h_1}\right] \quad (2.74)$$

The local error of approximation of the operator at (x,t) will be calculated; that is, $\partial^2 u/\partial x^2 - \bar{\delta}_x^2 u$ will be determined. With a Taylor-series expansion of the function $u(x,t)$ about the point (x,t),

$$u(x+h_2, t) = u(x,t) + h_2\frac{\partial u}{\partial x} + \frac{h_2^2}{2!}\frac{\partial^2 u}{\partial x^2} + \frac{h_2^3}{3!}\frac{\partial^3 u}{\partial x^3} + O(h_2^4)$$

$$u(x-h_1, t) = u(x,t) - h_1\frac{\partial u}{\partial x} + \frac{h_1^2}{2!}\frac{\partial^2 u}{\partial x^2} - \frac{h_1^3}{3!}\frac{\partial^3 u}{\partial x^3} + O(h_1^4)$$

we get

$$\bar{\delta}_x^2 u = \frac{\partial^2 u}{\partial x^2} + \frac{h_2 - h_1}{3}\frac{\partial^3 u}{\partial x^3} + O\left(\left(\frac{h_1+h_2}{2}\right)^2\right)$$

Thus, the operator $\bar{\delta}_x^2$, for an irregular grid, is of the first local order of approximation. Consequently, the introduction of a nonuniform grid may reduce the approximation order of a difference scheme.

Let us next consider some approximation methods for the Laplacian, which is frequently used in heat-transfer problems. Only the two-dimensional case will be treated, for $\nabla^2 u = \partial^2 u/\partial x^2 + \partial^2 u/\partial y^2$. A regular grid is assumed, such that $\Delta x = \Delta y = h$. The notation for the grid points is

BASIC CONCEPTS OF THE FINITE-DIFFERENCE METHOD

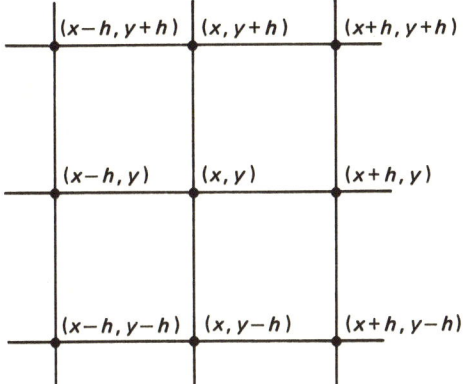

Fig. 2 Grid-point notation.

shown in Fig. 2. The difference analog of the Laplacian $\nabla^2 u$ will be denoted by $\nabla^2 u_h$ or $\nabla^2 u_{i,j}$. For easy reference, schematic diagrams of grid-point patterns used are given below.

In practice, approximations of the Laplacian using a five-point pattern

```
      *
  *   *   *
      *
```

of the form

$$\nabla^2 u_h = [u(x+h, y) + u(x-h, y) + u(x, y+h) + u(x, y-h) - 4u(x, y)]h^{-2}$$
$$= \nabla^2 u + \frac{h^2}{12}\left(\frac{\partial^4 u}{\partial x^4} + \frac{\partial^4 u}{\partial x^4}\right) + O(h^3) \tag{2.75}$$

are most frequently encountered. When the grid is turned by 45°, the formula obtained from the pattern

```
  *       *
      *
  *       *
```

is

$$\nabla^2 u_h = \frac{1}{2h^2}\left[u(x+h, y+h) + u(x+h, y-h) + u(x-h, y+h) + u(x-h, y-h)\right.$$
$$\left. - 4u(x, y)\right] = \nabla^2 u + \frac{h^2}{12}\left(\frac{\partial^4 u}{\partial x^4} + 6\frac{\partial^4 u}{\partial x^2 \partial y^2} + \frac{\partial^4 u}{\partial y^4}\right) + O(h^3) \tag{2.76}$$

Equation (2.75) is more accurate than Eq. (2.76). And, in some problems where numerical solutions are sought within a rectangular region, Eq. (2.75) does not necessarily require angular points, which introduce some uncertainty into the computational algorithm.

There are other means of approximating Laplacians. The formula obtained from a nine-point pattern

```
*  *  *
*  *  *
*  *  *
```

is widely used and is

$$\nabla^2 u_h = \frac{1}{6h^2} \{u(x+h, y+h) + u(x+h, y-h) + u(x-h, y+h)$$

$$+ u(x-h, y-h) + 4[u(x+h, y) + u(x-h, y) + u(x, y+h)$$

$$+ u(x, y-h)] - 20u(x, y)\} = \nabla^2 u + \frac{h^2}{12}\nabla^4 u + \frac{2h^4}{6!}$$

$$\cdot \left[\frac{\partial^4(\nabla^2 u)}{\partial x^4} + 4\frac{\partial^4(\nabla^2 u)}{\partial x^2 \partial y^2} + \frac{\partial^4(\nabla^2 u)}{\partial y^4}\right] + O(h^6) \quad (2.77)$$

Equation (2.77) is very convenient for the numerical solution of the Laplace equation or the Poisson equation with an analytical right-hand side. All the Laplacians on the right-hand side of (2.77) vanish, and the approximation order of the equation is $O(h^6)$. However, for solving more general problems, in which one of the terms is the Laplacian, or in numerical integration of the Poisson equation with a nonanalytical right-hand side, Eq. (2.77) is probably no better than Eq. (2.75).

Let us analyze some particular cases of Laplacian approximation using an irregular grid, a situation which may be encountered in the numerical integration of the Laplace equation in a region with a curvilinear boundary (Fig. 3). In the figure, boundary points are denoted with a cross, near-boundary points with a triangle, and other points, which will be called *interior* points, with a circle. It may be seen that, although the grid is uniform in the x and y directions, it is irregular near the boundary of the region. Possible approximations of the Laplacian for the typical situations shown in Fig. 4 are:

$$\nabla^2 u_h = \frac{2}{h_1 + h}\left(\frac{u_3 - u_0}{h} - \frac{u_0 - u_1}{h_1}\right) + \frac{u_2 - 2u_0 + u_4}{h^2} \quad (2.78a)$$

BASIC CONCEPTS OF THE FINITE-DIFFERENCE METHOD

$$\nabla^2 u_h = \frac{2}{h_1 + h}\left(\frac{u_3 - u_0}{h} - \frac{u_0 - u_1}{h_1}\right) + \frac{2}{h_2 + h}\left(\frac{u_2 - u_0}{h_2} - \frac{u_0 - u_4}{h}\right) \tag{2.78b}$$

$$\nabla^2 u_h = \frac{2}{h_1 + h_3}\left(\frac{u_3 - u_0}{h_3} - \frac{u_0 - u_1}{h_1}\right) + \frac{2}{h_2 + h}\left(\frac{u_2 - u_0}{h_2} - \frac{u_0 - u_4}{h}\right) \tag{2.78c}$$

The indices denote the numbers of the grid points at which the values of the functions are taken; they are the same as in Fig. 4.

This method of constructing difference schemes by substituting difference relations for the derivatives is not the best one, although it is the simplest. A more general method of composing difference algorithms involves approximation of the whole differential operator, rather than each of the derivatives separately. Let us illustrate this method by applying it to a Cauchy problem in a heat-conduction theory.

$$\frac{\partial u}{\partial t} - \frac{\partial^2 u}{\partial x^2} = 0 \quad u(x, 0) = \psi(x) \quad -\infty < x < \infty \quad t > 0 \tag{2.79}$$

To write a difference scheme, we take a set of grid points; i.e., we indicate the points at which the grid functions are to be related by difference equations. For the problem under consideration, for the grid $x_k = kh$ and $t^n = n\tau$, four grid points with the numbers (k, n), $(k+1, n)$, $(k-1, n)$, and $(k, n+1)$ are chosen as shown in Fig. 5.

The initial problem (2.79) is linear. Naturally, the difference

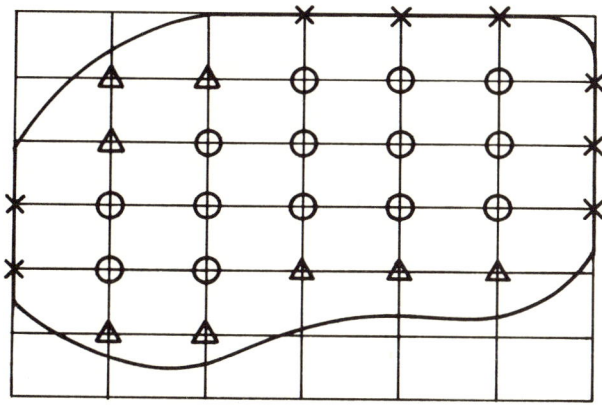

Fig. 3 Region with a curvilinear boundary.

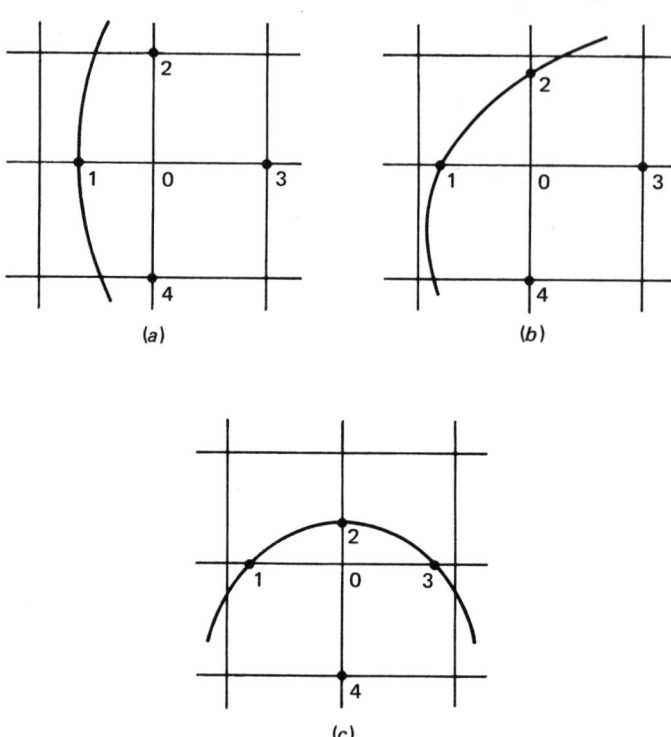

Fig. 4 Typical situations at curvilinear boundaries.

equations are to be written as linear relations. Such a relation between the functions at the points considered may be expressed in a general form as

$$U_k^{n+1} = a_{-1} U_{k-1}^n + a_0 U_k^n + a_1 U_{k+1}^n = \sum_{\alpha=-1}^{1} a_\alpha U_{k+\alpha}^n \qquad (2.80)$$

Assuming $U_k^0 = \psi(x_k)$, a difference problem is obtained with any combination of the a_α. The coefficients a_α are chosen such that problem (2.80) approximates the original problem (2.79) and is stable.

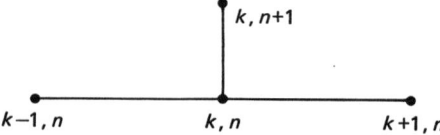

Fig. 5 The grid points.

BASIC CONCEPTS OF THE FINITE-DIFFERENCE METHOD

A spectral property will be utilized to study the stability. As is known, the necessary stability condition in this case has the form of the inequality

$$\left|\sum_{\alpha=-1}^{1} a_\alpha \exp(i\omega\alpha h)\right| \leqslant 1 \qquad (2.81)$$

Thus, we have one condition for the coefficients a_α.

Now consider the conditions for which the difference scheme approximates the original differential problem. To this end, exact solution (2.79) will now be expanded by a Taylor series with powers τ and h about the central point of the grid (x_k, t^n):

$$u_k^{n+1} = u_k^n + \tau \left(\frac{\partial u}{\partial t}\right)_k^n + \frac{\tau^2}{2}\left(\frac{\partial^2 u}{\partial t^2}\right)_k^n + \frac{\tau^3}{6}\left(\frac{\partial^3 u}{\partial t^3}\right)_k^n + \cdots$$

$$u_{k+\alpha}^n = u_k^n + \alpha h \left(\frac{\partial u}{\partial x}\right)_k^n + \frac{(\alpha h)^2}{2}\left(\frac{\partial^2 u}{\partial x^2}\right)_k^n + \frac{(\alpha h)^3}{6}\left(\frac{\partial^3 u}{\partial x^3}\right)_k^n + \cdots$$

Since the function $u(x, t)$ is the solution of Eq. (2.79), the following equalities hold:

$$\frac{\partial u}{\partial t} = \frac{\partial^2 u}{\partial x^2} \qquad \frac{\partial^2 u}{\partial t^2} = \frac{\partial^4 u}{\partial x^4} \qquad \frac{\partial^3 u}{\partial t^3} = \frac{\partial^6 u}{\partial x^6} \qquad \cdots$$

Using the above relations, we substitute the projection of the exact solution u_k^n on the grid points for the function U_k^n in (2.80). This gives

$$u_k^{n+1} - \sum_\alpha a_\alpha u_{k+\alpha} = \left(1 - \sum_\alpha a_\alpha\right) u_k^n - h \sum_\alpha \alpha a_\alpha \left(\frac{\partial u}{\partial x}\right)_k^n$$

$$+ \left(\tau - \frac{h^2}{2} \sum_\alpha \alpha^2 a_\alpha\right)\left(\frac{\partial^2 u}{\partial x^2}\right)_k^n - \frac{h^3}{6} \sum_\alpha \alpha^3 a_\alpha \left(\frac{\partial^3 u}{\partial x^3}\right)_k^n$$

$$+ \left(\frac{\tau^2}{2} - \frac{h^4}{24} \sum_\alpha a_\alpha \alpha^4\right)\left(\frac{\partial^4 u}{\partial x^4}\right)_k^n \cdots \qquad (2.82)$$

The order of approximation of scheme (2.80) obviously depends on that of the first nonzero terms of the series. Arbitrary and independent u_k^n,

$(\partial u/\partial x)_k^n$, $(\partial^2 u/\partial x^2)_k^n$, ... should be assumed. Taking the coefficients of the quantities as zero, we get a system of equations for α_i:

$$1 - \sum_\alpha a_\alpha = 0$$

$$\sum_\alpha \alpha a_\alpha = 0$$

$$\tau - \frac{h^2}{2} \sum_\alpha \alpha^2 a_\alpha = 0 \qquad (2.83)$$

$$\sum_\alpha \alpha^3 a_\alpha^2 = 0$$

. . . .

The greater the number of these equations that are satisfied, the higher is the order of approximation.

We have at our disposal only three indefinite coefficients, namely, a_{-1}, a_0, and a_1. Therefore, we may only hope to satisfy three of Eqs. (2.83). These are

$$a_1 + a_0 + a_{-1} = 1 \qquad a_1 - a_{-1} = 0 \qquad a_1 + a_{-1} = \frac{2\tau}{h^2}$$

We get

$$a_0 = 1 - \frac{2\tau}{h^2} \qquad a_{-1} = a_1 = \frac{\tau}{h^2} \qquad (2.84)$$

With these values for a_α, the first nonzero terms of series (2.82) are of the order $O(\tau^2 + h^2)$. Since a difference scheme is usually written as $(1/\tau)(U_k^{n+1} - \Sigma_\alpha a_\alpha U_{k+\alpha}^n)$, the order of approximation of scheme (2.80) is determined by $O(\tau + h^2)$.

Further, we are to satisfy stability condition (2.81). Substitution of (2.84) into (2.81) gives the inequality

$$\left| \sum_\alpha a_\alpha \exp(i\omega\alpha h) \right| \leq \left| 1 + \frac{2\tau}{h^2}(\cos \omega h - 1) \right|$$

BASIC CONCEPTS OF THE FINITE-DIFFERENCE METHOD

which is satisfied at $\tau \leqslant h^2/2$, the stability condition of difference problem (2.80).

This method of constructing the computational equations may be applied to any other problem. First of all, from heuristic considerations, a pattern of grid points and a form of difference relations containing indefinite coefficients are chosen. With the approximation and stability conditions satisfied, the problem is then reduced to the solution of an algebraic system of equations and inequalities.

Next, let us describe one more approach to the construction of finite-difference schemes. First, however, some general remarks will be made.

Various physical processes, such as heat conduction, diffusion, and oscillations, are governed by certain conservation laws (heat, mass, momentum, etc.). The derivations of the equations of mathematical physics are usually based on certain integral conservation relations applied to a small volume. A differential equation is obtained from a balance equation with vanishing volume and with the assumption of the existence of continuous derivatives.

The meaning of the finite-difference technique stems from the transition from a continuum to a certain discontinuous model. All the properties of the physical process should be preserved. The conservation laws pertain primarily to such properties. Difference schemes that express the conservation laws on a grid are called *conservational* schemes. For conservational schemes, the conservation laws should hold within the whole grid region as an algebraic consequence of the difference equations.

Conservational difference schemes must be based on balance equations written for meshes of the grid region. The integrals and derivatives entering into the balance equations should be replaced by approximate difference expressions. This procedure, used for conservational difference schemes, is called the *integrointerpolation* method (the *balance* method) [6].

The integrointerpolation method will be illustrated with the solution of a steady temperature distribution in a homogeneous rod of unit length $(0 \leqslant x \leqslant 1)$:

$$\frac{d}{dx}\left[\kappa(x)\frac{du}{dx}\right] - q(x)u(x) + f(x) = 0 \quad u(0) = \psi_1 \quad u(1) = \psi_2 \quad (2.85)$$

where $\kappa(x) > 0$ is the thermal conductivity, $q(x) \geqslant 0$ is the strength of the sinks [or sources if $q(x) < 0$] proportional to the temperature, and $f(x)$ is the distribution density of the external heat sources (or sinks).

The interval $0 \leq x \leq 1$ is divided into equal steps by points $x_k = kh$ ($k = 0, 1, 2, \ldots, K$), and the balance equation for the unit length l_k ($x_{k-1/2} \leq x \leq x_{k+1/2}$) of the section is written as

$$Q(x_{k-1/2}) - Q(x_{k+1/2}) - \int_{l_k} q(x)u(x)\,dx + \int_{l_k} f(x)\,dx = 0 \qquad (2.86)$$

where $Q(x) = -\kappa(x)\,du/dx$ is the heat flux transferred through cross section x. Heat is assumed to be supplied from $x_{k-1/2}$ and removed through $x_{k+1/2}$. The third term in equality (2.86) is the quantity of heat transferred to the environment at the lateral surface of the rod. The last integral is the heat generated over the length l_k by heat sources distributed with density $f(x)$.

A difference equation is obtained by substituting, for the integral containing $u(x)$, a linear combination of $u(x)$ at the grid points. The simplest interpolation considers $u(x)$ constant over the entire length l_k, so that $u(x) = u(x_k) = \text{const}$. Then

$$\int_{l_k} q(x)u(x)\,dx \approx h\bar{q}_x u_k \qquad (2.87)$$

where $\bar{q}_k$ is the mean value of $q(x)$ over the integration region, a particular form of which may be written as follows:

$$\bar{q}_k = \frac{1}{h}\int_{l_k} q(x)\,dx$$

To express the heat fluxes $Q(x_{k-1/2})$ and $Q(x_{k+1/2})$ in terms of a combination of functions $u(x)$ at the grid points, the equality

$$-\frac{Q(x)}{\kappa(x)} = \frac{du}{dx}$$

is integrated over the intervals $x_{k-1} \leq x \leq x_k$ and $x_k \leq x \leq x_{k+1}$. The function $Q(x)$ will be assumed constant over the intervals and equal to $Q(x_{k-1/2})$ and $Q(x_{k+1/2})$. The integration results in

BASIC CONCEPTS OF THE FINITE-DIFFERENCE METHOD

$$u_{k-1} - u_k = Q_{k-1/2} \int_{x_{k-1}}^{x_k} \frac{dx}{\kappa(x)} \qquad u_k - u_{k+1} = Q_{k+1/2} \int_{x_k}^{x_{k+1}} \frac{dx}{\kappa(x)}$$

With the notations

$$a_k = \left(\frac{1}{h} \int_{x_{k-1}}^{x_k} \frac{dx}{\kappa(x)}\right)^{-1} \qquad a_{k+1} = \left(\frac{1}{h} \int_{x_k}^{x_{k+1}} \frac{dx}{\kappa(x)}\right)^{-1}$$

we get

$$Q_{k-1/2} = -a_k \frac{u_k - u_{k-1}}{h} \qquad Q_{k+1/2} = -a_{k+1} \frac{u_{k+1} - u_k}{h} \tag{2.88}$$

Finally, consider the mean distribution density of the external sources over the length l_k:

$$\bar{f}_k = \frac{1}{h} \int_{l_k} f(x)\, dx \tag{2.89}$$

The substitution of (2.87) through (2.89) into the balance equation (2.86) gives the conservational difference scheme:

$$\frac{1}{h}\left(a_{k+1} \frac{U_{k+1} - U_k}{h} - a_k \frac{U_k - U_{k-1}}{h}\right) - \bar{q}_k U_k = -\bar{f}_k \tag{2.90}$$

This difference equation applies for the fixed grid point x_k. Since x_k is arbitrary, Eq. (2.90) is valid for all internal grid points. Since the coefficients of the equation are defined at all the grid points by the same equations, scheme (2.90) is homogeneous. It may easily be inferred that the scheme approximates the original problem (2.85) with order $O(h)$.

The difference scheme is indeed a conservative scheme; i.e., the heat-conservation law holds over the whole grid region as an algebraic consequence of equation (2.90). Using $\bar{Q}_{k-1/2} = -a_k(U_k - U_{k-1})/h$ to denote the difference expression for the heat flux through the cross section $x = x_{k-1/2}$, we may write equality (2.90) as

$$\bar{Q}_{k-1/2} - \bar{Q}_{k+1/2} + h\bar{f}_k = h\bar{q}_k U_k$$

Summation over $k = 1, 2, \ldots, K - 1$ gives the difference heat-conservation law for the whole grid region:

$$\bar{Q}_{1/2} - \bar{Q}_{K-1/2} + h \sum_{k=1}^{K-1} f_k = h \sum_{k=1}^{K-1} \bar{q}_k U_k \qquad (2.91)$$

By comparison of (2.91) and (2.86), one draws the conclusion that expression (2.91) is the difference approximation of the integral conservation law (2.86).

Different approaches to finite-difference schemes are available. The most promising are probably the procedures that produce difference schemes with a high order of approximation, since they increase the accuracy of the resultant numerical solution through more accurate approximation of the original problem via the assumption of a smooth solution, rather than with smaller grid steps. Among the procedures most widely used are those based on the variational methods of Ritz and Galerkin and the least-squares method. It should, however, be emphasized that the use of these methods makes the numerical algorithms more complicated and requires profound knowledge of variational theory. Since the aim of this book is to present finite-difference theory in a simple and comprehensive form, these methods of constructing difference schemes will not be considered in detail.

2.7 SOLUTION OF DIFFERENCE EQUATIONS

It was demonstrated above that, for a given differential problem, a large number of finite-difference schemes are possible. Explicit schemes produce the simplest algorithms. In such a case, the difference scheme itself is the equation that is used to solve explicitly for the nth time (iterational) level in terms of the known solutions at previous levels. This is why such schemes are called explicit.

Explicit methods are, unfortunately, stable only for certain relations between the time and space sets of the grid ($\tau \leq h^2$). The stability condition therefore requires very small time steps, thus making the computation time prohibitively large.

Implicit finite-difference schemes are free of this disadvantage. They permit an independent choice of the time and space parameters of the grid. In this case, however, solution of the difference problem requires a system

BASIC CONCEPTS OF THE FINITE-DIFFERENCE METHOD

of algebraic equations. Therefore, after the difference boundary-value problem which approximates the original differential system is formulated, a convenient solution method should be indicated.

The numbers of equations and unknowns are proportional to h^{-p}, where p is the number of space variables. If the peculiarities of the system are neglected and it is solved as a general kind of system, this requires a tremendous number of arithmetic operations. A peculiarity of difference equations is that the matrix of the coefficients is of a tridiagonal structure that allows successful use of some special methods.

One of the most widely used methods for difference equations is the Gaussian two-step elimination method, or the factorization procedure [23]. Consider a system of difference equations of the form

$$A_k U_{k-1} + B_k U_k + C_k U_{k+1} = F_k \quad k = 1, 2, 3, \ldots, K-1 \quad (2.92)$$

with the prescribed boundary conditions

$$U_0 = a_0 U_1 + b_0 \quad U_K = a_K U_{K-1} + b_K \quad (2.93)$$

where A_k, B_k, C_k, F_k, a_0, b_0, a_K, and b_K are known quantities. All implicit two-level difference problems with one space variable and boundary conditions of the first, second, and third kinds are of this form. It will be demonstrated below how, on the basis of the fractional-step principle, finite-difference schemes for multidimensional problems may be reduced to the form of (2.92) and (2.93).

The solution of the system (2.92)-(2.93) will be sought in a form consistent with the boundary conditions

$$U_k = \alpha_k U_{k+1} + \beta_k \quad k = 1, 2, 3, \ldots, K-1 \quad (2.94)$$

where α_0 and β_0 are unknown coefficients. Equation (2.94) is valid at all indices k. Therefore,

$$U_{k-1} = \alpha_{k-1} U_k + \beta_{k-1}$$

Substitution of this expression into (2.92) yields

$$A_k(\alpha_{k-1} U_k + \beta_{k-1}) + B_k U_k + C_k U_{k+1} = F_k$$

The above equation may be rewritten in the form (2.94):

$$U_k = -\frac{C_k}{A_k \alpha_{k-1} + B_k} U_{k+1} + \frac{F_k - A_k \beta_{k-1}}{A_k \alpha_{k-1} + B_k} \qquad (2.95)$$

A comparison of expressions (2.95) and (2.94) gives recurrent relations for α_k and β_k:

$$\alpha_k = -\frac{C_k}{A_k \alpha_{k-1} + B_k} \qquad \beta_k = \frac{F_k - A_k \beta_{k-1}}{A_k \alpha_{k-1} + B_k} \qquad (2.96)$$

But $\alpha_0 = a_0$ and $\beta_0 = b_0$ are known from the boundary condition at a point with index $k = 0$. Equations (2.96) allow successive calculation of α_1, $\alpha_2, \alpha_3, \ldots, \alpha_{K-1}$ and $\beta_1, \beta_2, \beta_3, \ldots, \beta_{K-1}$.

Then, from (2.94), for the point with index $k = K - 1$, we may write

$$U_{K-1} = \alpha_{K-1} U_K + \beta_{K-1} \qquad (2.97)$$

This relation should hold, irrespective of the second boundary condition that relates the same values. Simultaneous solution of (2.93) and (2.97) gives

$$U_K = (a_K \beta_{K-1} + b_K)(1 - a_K \alpha_{K-1})^{-1}$$

With U_K known, recurrence formula (2.94) allows successive determination of $U_{K-1}, U_{K-2}, \ldots, U_2, U_1$ in the reverse direction; thus, difference problem (2.92)–(2.93) is solved.

For convenience in some problems, the computations may be arranged so that the U_k are found in succession, beginning from the left-hand boundary. Then the computation equations are derived in a similar way as

$$\begin{aligned} \xi_k &= -\frac{A_k}{C_k \xi_{k+1} + B_k} & \eta_k &= \frac{F_k - C_k \eta_{k+1}}{C_k \xi_{k+1} + B_k} \\ \xi_K &= a_K & \eta_K &= b_K \\ U_0 &= \frac{a_0 \eta_1 + b_0}{1 - a_0 \xi_1} & U_k &= \xi_k U_{k+1} + \eta_k \end{aligned} \qquad (2.98)$$

BASIC CONCEPTS OF THE FINITE-DIFFERENCE METHOD

The stability of factorization formulas (2.94) and (2.96) evidently requires that the coefficient a does not exceed unity. Then, the truncation errors from formula (2.94) do not increase.

The stability of these equations is provided [23] by

$$A_k > 0 \quad C_k > 0 \quad -B_k \geqslant A_k + C_k \quad 0 \leqslant a_0 < 1 \quad 0 \leqslant a_K < 1 \quad (2.99)$$

Indeed, $\alpha_0 = a_0 < 1$, and if $0 \leqslant \alpha_{k-1} < 1$, then

$$0 < \alpha_k = -\frac{C_k}{B_k + A_k \alpha_{k-1}} = \frac{C_k}{C_k + (1 - \alpha_{k-1})A_k - A_k - B_k - C_k} < 1$$

In some cases, the limitations on a may be reduced [6]. For example, the computational equations remain valid if (2.99) is replaced with the conditions

$$A_k > 0 \quad C_k > 0 \quad -B_k \geqslant A_k + C_k$$
$$0 \leqslant a_0 \leqslant 1 \quad 0 \leqslant a_K \leqslant 1 \quad -B_k \equiv A_k + C_k \quad (2.100)$$

or

$$A_k > 0 \quad C_k > 0 \quad -B_k \geqslant A_k + C_k$$
$$0 \leqslant a_0 \leqslant 1 \quad 0 \leqslant a_K \leqslant 1 \quad a_0 + a_K \leqslant 1 \quad (2.101)$$

The factorization algorithm is simple and effective. Its usage increases the number of operations for one time level approximately by a factor of 5, compared with explicit schemes. Thus, implicit schemes are advantageous if the time step τ for the required accuracy is as large as five times that for the stability requirement in an appropriate explicit scheme.

A similar method exists for multidimensional problems. The initial system of difference equations may be written in the form (2.92). Then the coefficients A_k, B_k, and C_k are not scalar quantities, but rather are matrices. Accordingly, U_k and F_k are vectors. The formulas for the solution of such a system may be obtained as above. The procedure is then called *matrix factorization*. It is rarely used, because very large amounts of computer storage are required by the large number of matrices of arbitrary structure.

Iteration methods are also frequently used to solve difference-equation systems. Before we consider these methods, some general comments are offered.

In the general case, a system of linear difference equations may be written in a vectorial matrix form as

$$A\mathbf{u} = b \tag{2.102}$$

where $\mathbf{u}$ is the vector of the unknowns. The matrix of coefficients A is determined by finite-difference approximation and a discrete grid, and b by the boundary conditions and the right-hand side. The matrix A is not *nonsingular*, and that allows $\mathbf{u}$ to be written as $\mathbf{u} = A^{-1}b$. For an approximate solution of (2.102) by the iteration method, we write

$$\mathbf{u} = B\mathbf{u} + f \tag{2.103}$$

which is always possible by the use of very simple transformations. Linear iteration of (2.103) may be described by

$$\mathbf{u}^{s+1} = B\mathbf{u}^s + f \tag{2.104}$$

where $\mathbf{u}^0$ is the initial approximation. If successive $\{\mathbf{u}^s\}$'s are convergent, then they converge to the solution of the linear system

$$[E - B]\mathbf{u} = f \tag{2.105}$$

where E is a unit matrix.

In order that the limit of successive $\{\mathbf{u}^s\}$'s be the solution of (2.102), it is necessary that a *nonsingular* matrix C exist for which

$$C[E - B] = A \qquad Cf = b$$

Many such matrices may exist. Since each iteration requires one multiplication of the vector by matrix B, a matrix C is chosen so that the corresponding matrix B is weakly satisfied.

To determine the convergence condition of the iteration process (2.104), the solution error is defined by

$$\epsilon^s = \mathbf{u} - \mathbf{u}^s$$

With (2.104) and (2.105), we get

$$\epsilon^{s+1} = \mathbf{u} - \mathbf{u}^{s+1} = \mathbf{u} + B\mathbf{u}^s - f = \mathbf{u} - B\mathbf{u}^s - [E - B]\mathbf{u}$$

BASIC CONCEPTS OF THE FINITE-DIFFERENCE METHOD

and

$$\epsilon^{s+1} = B\epsilon^s = B^s \epsilon^0 \tag{2.106}$$

Equation (2.106) implies that the iteration process (2.104) converges at all ϵ^0 only when all the eigenvalues of matrix B have moduli less than unity. Spectral radius $\rho(B)$ of matrix B is defined as

$$\rho(B) = \max_m |\lambda_m|$$

where λ_m is the mth eigenvalue of matrix B. The convergence condition for the iteration process (2.104) is then the inequality

$$\rho(B) < 1 \tag{2.107}$$

Let us consider some iteration procedures that are widely used for difference equations. As an example, consider the Laplace equation

$$\frac{\partial^2 u}{\partial x^2} + \frac{\partial^2 u}{\partial y^2} = 0 \tag{2.108}$$

determined in the square region $0 \leq x \leq 1$, $0 \leq y \leq 1$, with $u(x, y)$ prescribed at all the sides. The finite-difference scheme is constructed as follows:

$$\frac{u_{k+1,l} - 2u_{k,l} + u_{k-1,l}}{h^2} + \frac{u_{k,l+1} - 2u_{k,l} + u_{k,l-1}}{h^2} = 0 \tag{2.109}$$

The functions $u_{0,l}$, $u_{K,l}$, $u_{k,0}$, $u_{k,L}$ are prescribed.

To write the iteration scheme, the difference equation (2.109) is reduced to the form of (2.103),

$$u_{k,l} = \frac{1}{4}(u_{k+1,l} + u_{k-1,l} + u_{k,l+1} + u_{k,l-1})$$

from which the simple iteration equation

$$u_{k,l}^{s+1} = \frac{1}{4}(u_{k+1,l}^s + u_{k-1,l}^s + u_{k,l+1}^s + u_{k,l-1}^s) \tag{2.110}$$

immediately follows. Every $u_{k,l}$ is computed irrespective of the others. Procedure (2.110) is called a *simple* iteration method. It is sometimes called the *Jacobi* iteration method.

The eigenvalues of matrix B for procedure (2.110) may be demonstrated [8] to equal $\lambda_m = \cos m\pi h$, and $\rho(B) = \cos \pi h$. Thus, the iteration procedure (2.110) is always convergent. With a small grid step, $\cos \pi h \approx 1 - 0.5(\pi h)^2$. That implies that the iteration procedure converges extremely slowly in this case.

Besides slow convergence, the simple iteration method has one additional disadvantage: The new values of $u_{k,l}^{s+1}$ are used only after all the iterations at the step $s + 1$ have been completed. If the new values of $u_{k,l}^{s+1}$ are utilized directly as soon as they are computed, then

$$u_{k,l}^{s+1} = \frac{1}{4}(u_{k+1,l}^{s} + u_{k-1,l}^{s+1} + u_{k,l+1}^{s} + u_{k,l-1}^{s+1}) \tag{2.111}$$

and the method is known as the *Saidel* iteration method.

The rate of convergence of the Saidel method is somewhat higher $[\rho(B) = \cos^2 \pi h]$ than that of the Jacobi iteration method. In addition, it requires $(K + 1)(L + 1)$ fewer storage locations, since the method always replaces the old $u_{k,l}^{s+1}$ with the new ones.

The rate of convergence of the above iteration methods may be increased by the introduction of a parameter that speeds up the iteration procedure. The obvious identity $u_{k,l}^{s+1} \equiv u_{k,l}^{s} + (u_{k,l}^{s+1} - u_{k,l}^{s})$ implies that a new value of $u_{k,l}^{s}$ is obtained by the addition of $u_{k,l}^{s+1} - u_{k,l}^{s}$ to the old value of $u_{k,l}^{s}$. Variations of the addend permit some control of the iteration process. The substitution of some controlling factor γ into the above identity before the parentheses and using (2.111) in place of $u_{k,l}^{s+1}$ give

$$u_{k,l}^{s+1} = (1 - \gamma)u_{k,l}^{s} + \frac{\gamma}{4}(u_{k-1,l}^{s+1} + u_{k+1,l}^{s} + u_{k,l-1}^{s+1} + u_{k,l+1}^{s}) \tag{2.112}$$

Theoretical considerations based on analysis of the eigenvalues of the matrix of coefficients in (2.112) reveal [9] that the resultant iteration procedure converges at all γ consistent with $0 < \gamma < 2$. The highest rate of convergence is found at a certain optimum $\gamma = \gamma_0$ where $1 < \gamma_0 < 2$. Since the value of the parameter γ is normally greater than unity, the iteration process is sometimes called the *overrelaxation method*, and γ the *relaxation*

BASIC CONCEPTS OF THE FINITE-DIFFERENCE METHOD

parameter. This is a very useful method, and computational experience shows that at γ close to γ_0, its rate of convergence exceeds that of the Saidel method by more than an order of magnitude.

Unfortunately, the determination of the optimal relaxation parameter is not always easy. In [9] it is demonstrated that the optimum γ_0 should be consistent with the inequality

$$\gamma_0^2 \nu^2 = 4(\gamma_0 - 1) \tag{2.113}$$

where ν is the largest eigenvalue of the matrix composed of the coefficients of the appropriate simple iteration. But it is sometimes different to determine ν as well.

There are several practical ways of determining γ_0. The most simple and widely used involves the selection of an approximate γ_0 from numerical experiments. For example, starting with $\gamma = 1$ (the Saidel method) and carefully watching the convergence rate of the iteration procedure, continually increase the value of γ. As soon as the rate of convergence begins decreasing, no further increase of γ is necessary. The value of γ at that time should be taken as γ_0.

An approximate γ_0 is sometimes found from Eq. (2.113). It may easily be seen that the smaller root of the equation in the range of $0 < \gamma_0 < 2$ is given by

$$\gamma_0 = \frac{2}{1 + \sqrt{1 - \nu^2}} \tag{2.114}$$

In practical situations, ν is often determined roughly by several simple iterations, and substitution of the ν thus into (2.114) gives an approximate γ_0.

For the case in which scheme (2.112) is used for the solution of the Laplace equation, the optimum relaxation parameter γ_0 may easily be found, since the value of ν is known ($\nu = \cos \pi h$). Then

$$\gamma_0 = \frac{2}{1 + \sin \pi h} \tag{2.115}$$

The same γ_0 is also valid for the relaxation iteration procedure applied to the Poisson equations

$$\frac{\partial^2 u}{\partial x^2} + \frac{\partial^2 u}{\partial y^2} = f(x, y)$$

For some difference equations, procedures known as *block-type* iteration procedures are sometimes used. The idea here is based on the fact that, at every iteration step, the solution of a general system of difference equations is reduced to the solution of several systems of a lower order. Although their rate of convergence is somewhat higher than that of point-type iteration methods, block-type iteration methods require twice as many storage locations. Therefore, block-type iteration method possesses noticeable advantages over its point-type counterpart.

Iteration procedures based on the fractional-step method seem highly promising; they reduce the solution of difference equations for multi-dimensional problems to that of one-dimensional types [22]. A description of the procedures will be presented after a discussion of the fractional-step method.

There are other effective methods for difference equations, but most of them involve complicated numerical algorithms and a high mathematical level. At present, when computer operating times are expressed in microseconds and high-speed memory is not a problem, the simplicity of application of numerical methods is of primary importance.

Chapter 3

Heat-Conduction Problems

3.1 SIMPLE HEAT-CONDUCTION PROBLEMS

Heat propagation in a medium at rest is governed by a heat-conduction equation [10] of the general form

$$c\rho \frac{\partial T}{\partial t} = \nabla \kappa \nabla T + Q \tag{3.1}$$

where $T(\bar{x}, t)$ = temperature
c = heat capacity of a unit mass
ρ = density
κ = thermal conductivity
$Q(\bar{x}, t)$ = heat-source density or heat quantity generated per unit time in a unit volume

$\bar{x}$ = vector of location of a point in space
t = time

The thermal conductivity and heat capacity may depend on $\bar{x}$, t, and T.

Equation (3.1), with suitable boundary and initial conditions, represents the problem of the temperature distribution at any time and at every point of the volume, under consideration. The initial temperature distribution is known; the temperature behavior at the boundaries is prescribed; the heat fluxes (if any) are known; the temperature distributions at subsequent times are to be found.

Before describing the methods used to solve such general three-dimensional problems, we shall consider some simpler cases frequently encountered in practice.

One such problem is to solve for the temperature distribution in an insulated thin rod of finite length. With the temperatures at the rod ends prescribed, a known temperature distribution inside the rod, say at $t = 0$, and heat sources in the rod (if any) described, the temperature distribution at any time t and at every point of the rod may be found from Eq. (3.1), which becomes, for this case,

$$c\rho \frac{\partial T}{\partial t} = \frac{\partial}{\partial x}\left(\kappa \frac{\partial T}{\partial x}\right) + Q \tag{3.2}$$

Here, a coordinate system is chosen so that the x axis is directed along the rod.

It may easily be seen that Eq. (3.2) may also be used to describe the temperature distribution in a thick plate whose surfaces have equal temperatures, in the direction normal to the plate.

Equation (3.2) becomes especially simple if $c\rho$ and κ are constant. That may be the case when the rod (plate) is made of a homogeneous material whose thermophysical properties change slightly over the time and temperature ranges under consideration. In such a case, the equation is written as

$$\frac{\partial T}{\partial t} = a \frac{\partial^2 T}{\partial x^2} + \tilde{f} \tag{3.3}$$

where $a = \kappa/c\rho$ is the thermal diffusivity, and $\tilde{f} = Q/c\rho$.

For computational convenience, the initial differential problem is ordinarily formulated in a dimensionless form. The quantity $\theta = (T - T_0)/\Delta T$ may be used as a dimensionless temperature, where T_0 is a fixed

HEAT-CONDUCTION PROBLEMS

temperature (generally, the ambient temperature), and ΔT is a characteristic temperature difference. If the temperature range of the problem is known, then $\theta = (T - T_{\min})/(T_{\max} - T_{\min})$ is frequently taken as the dimensionless temperature. In this case, θ ranges from zero to unity.

If the rod length is l and the solution of Eq. (3.3) is desired over the length $0 \leqslant x \leqslant l$, then with the introduction of the new variables $x' = x/l$ and $t' = t/l^2$, the region of the problem may be reduced to the interval $0 \leqslant x' \leqslant 1$. Equation (3.3) may now be written as

$$\frac{\partial \theta}{\partial t} = a \frac{\partial^2 \theta}{\partial x^2} + f \qquad f = l^2 \tilde{f} \tag{3.4}$$

where the primes have been omitted from the new independent variables for convenience.

For the completeness, the differential equation (3.4) must be supplemented with boundary and initial conditions; i.e., the temperature distribution along the rod at the time $t = 0$ and the temperatures at points $x = 0$ and $x = 1$ (at the ends) must be specified.

At each of the boundaries, the temperature (the first-kind boundary condition), or the heat flux (the second-kind condition), or a combination of these quantities (the boundary condition of the third kind) may be given. In the general case, the boundary conditions may be written as

$$\alpha \frac{\partial \theta}{\partial x} + \beta \theta + \gamma = 0$$

Thus, when $\alpha = 0$ and $\beta \neq 0$, the boundary condition is of the first kind; $\alpha \neq 0$ and $\beta = 0$ is a second-kind boundary condition; and $\alpha \neq 0$ and $\beta \neq 0$ is a third-kind condition.

Only difference methods for one-dimensional heat-conduction problems with constant coefficients will be treated here. For simplicity, first-kind boundary conditions are assumed at $x = 0$ and $x = 1$:

$$\theta(x, 0) = \psi(x) \qquad \theta(0, t) = \psi_1(t) \qquad \theta(1, t) = \psi_2(t) \tag{3.5}$$

The extension of the methods to boundary conditions of the second and third kinds is not generally difficult and will be considered later.

3.2 FINITE-DIFFERENCE ALGORITHMS FOR ONE-DIMENSIONAL HEAT-CONDUCTION PROBLEMS WITH CONSTANT THERMAL DIFFUSIVITIES

Let us consider the heat-conduction problem (3.4)–(3.5) of the previous section. We shall obtain the solution by a finite-difference technique. To introduce a uniform differencing grid, the interval $0 \leq x \leq 1$ is divided into J equal parts by $x_i = ih$, $i = 0, 1, 2, \ldots, J$, with $h = 1/J$. The solution is sought at the points x_i for time steps $t^n = n\tau$, $n = 0, 1, 2, \ldots$.

It should be pointed out that it is impossible to recommend one best finite-difference algorithm. Each of the finite-difference procedures possesses its own advantages and disadvantages. The choice of a particular procedure depends on such factors as the characteristics of the problem, the capabilities of the available computer, and the experience of the individual.

The simplest and most natural method for problem (3.4)–(3.5) is based on an explicit approximation of the original differential equation:

$$\frac{\theta_i^{n+1} - \theta_i^n}{\tau} = \frac{a}{h^2}(\theta_{i+1}^n - 2\theta_i^n + \theta_{i-1}^n) + f_i^n \qquad (3.6)$$

$$\theta_i^0 = \psi_i \qquad \theta_0^n = \psi_1^n \qquad \theta_J^n = \psi_2^n$$

The algorithm for the numerical solution is expressed by

$$\theta_i^{n+1} = \zeta(\theta_{i+1}^n - \theta_{i-1}^n) + (1 - 2\zeta)\theta_i^n + \tau f_i^n \qquad \zeta = \frac{a\tau}{h^2}$$

With θ_i^0 known, successive determination of θ_i^1, $\theta_i^2, \ldots$ is not difficult.

Difference schemes of this type were considered earlier. It was shown that computational stability requires the time steps to be conformable with

$$\tau \leq \frac{h^2}{2a} \qquad (3.7)$$

Substitution of the exact solution into (3.6), followed by Taylor-series expansion of $\theta(x_i, t^n + \tau)$, $\theta(x_i + h, t^n)$, $\theta(x_i - h, t^n)$ about the point (x_i, t^n) readily gives the error expression for the method:

$$R(x, t) = \frac{\tau}{2}\frac{\partial^2 \theta}{\partial t^2} - \frac{ah^2}{12}\frac{\partial^4 \theta}{\partial x^4} + O(\tau^2 + h^4)$$

The error is thus of the order $O(\tau + h^2)$.

HEAT-CONDUCTION PROBLEMS

It may easily be demonstrated that one may select τ such that the order of the error is $O(h^4)$. Indeed, because of the form of differential equation (3.4), we may write

$$\frac{\partial^2 \theta}{\partial t^2} = a^2 \frac{\partial^4 \theta}{\partial x^4} + \frac{\partial^2 f}{\partial x^2}$$

Substitution of this expression into the error formula gives

$$R(x, t) = a\left(\frac{a\tau}{2} - \frac{h^2}{12}\right) \frac{\partial^4 \theta}{\partial x^4} + O(\tau^2 + h^4)$$

which implies that if $\tau = h^2/6a$ is chosen and in the right-hand side of Eq. (3.6) f_i^n is replaced by

$$f_i^n + \frac{\tau}{2}\left(\frac{\partial^2 f}{\partial x^2}\right)_i^n$$

then $R \approx O(h^4)$ holds. The order of approximation is obviously is not disturbed upon the substitution

$$f_i^n + \frac{\tau}{2}\left(\frac{\partial^2 f}{\partial x^2}\right)_i^n = f_i^n + \frac{1}{12a}(f_{i+1}^n - 2f_i^n + f_{i-1}^n)$$

Since stability condition (3.7) is satisfied with $\tau = h^2/6a$, the higher-accuracy scheme is stable.

Generally speaking, the stability condition for difference schemes of this type is rather cumbersome, since it requires very small time steps and a large amount of computation time. Therefore, in spite of the simplicity and convenience of the algorithms for explicit schemes, they fail in many cases. Nevertheless, these schemes are often used in practice.

More promising are implicit schemes, which allow an independent choice of grid parameters. Moreover, very effective numerical methods are available for one-dimensional problems.

Consider a single-parameter family of difference schemes for problem (3.4)-(3.5),

$$\frac{\theta_i^{n+1} - \theta_i^n}{\tau} = \frac{\sigma a}{h^2}(\theta_{i+1}^{n+1} - 2\theta_i^{n+1} + \theta_{i-1}^{n+1}) + \frac{a(1-\sigma)}{h^2}$$

$$\cdot (\theta_{i+1}^n - 2\theta_i^n + \theta_{i-1}^n) + \varphi_i^n \quad (3.8)$$

$$\theta_i^0 = \psi_i \qquad \theta_0^n = \psi_1^n \qquad \theta_J^n = \psi_2^n$$

where φ_i^n is the grid function approximating the right-hand side of (3.4), for example, $\varphi_i^n = f(x_i, t^n + 0.5\tau) = f_i^{n+1/2}$ and $\sigma > 0$ is an arbitary real parameter. By fixing various values of σ, it is possible to obtain difference schemes with different properties. In particular, for $\sigma = 0$, the difference scheme (3.8) reduces to scheme (3.6), which was discussed previously.

The family of difference schemes (3.8) is sometimes called a *weighted* scheme for a heat-conduction equation. With $\sigma \neq 0$, the difference scheme (3.8) is implicit.

This case will next be considered in more detail. To find θ_i^{n+1} from such a scheme, it is necessary to solve the system of algebraic equations

$$\theta_{i-1}^{n+1} - \left(2 + \frac{h^2}{a\sigma\tau}\right)\theta_i^{n+1} + \theta_{i+1}^{n+1} = -F_i^n \quad i = 1, 2, 3, \ldots, J-1 \quad (3.9)$$

where

$$F_i^n = \frac{1-\sigma}{\sigma}(\theta_{i-1}^n - 2\theta_i^n + \theta_{i+1}^n) + \frac{h^2}{a\sigma\tau}\theta_i^n + \frac{h^2}{\sigma a}\varphi_i^n$$

The boundary values $\theta_0^n = \psi_1^n$ and $\theta_J^n = \psi_2^n$ are known, as well as the functions θ_i^n at all the points x_i, at the nth time step.

The factorization method discussed in Sec. 2.7 is most useful for algebraic systems of this kind. First, coefficients $\alpha_1, \alpha_2, \ldots, \alpha_{J-1}$ and $\beta_1, \beta_2, \ldots, \beta_{J-1}$ must be determined. Recurrent formulas for this case are

$$\alpha_i = \left(2 + \frac{h^2}{a\sigma\tau} - \alpha_{i-1}\right)^{-1} \quad \beta_i = (\beta_{i-1} + F_i^n)\alpha_i$$

Since $\theta_0^n = \psi_1^n$ at $x = 0$, we have $\alpha_0 = 0$ and $\beta_0 = \psi_1^n$ at the boundary. After the coefficients α_i and β_i are calculated, successive determination of $\theta_{J-1}^{n+1}, \theta_{J-2}^{n+1}, \ldots, \theta_1^{n+1}$ is possible from the formula

$$\theta_i^{n+1} = \alpha_i \theta_{i+1}^{n+1} + \beta_i$$

with $\theta_J^{n+1} = \psi_2^{n+1}$ known. All θ_i^{n+1} at the $(n+1)$th time step are thus found, and we may proceed to the calculation of θ_i^{n+2} at the $(n+2)$th step.

Simple calculations show that the difference scheme (3.8) is of order $O(\tau + h^2)$ at any σ. With $\sigma = 0.5$, the Crank-Nickolson scheme is obtained,

HEAT-CONDUCTION PROBLEMS

and it is of order $O(\tau^2 + h^2)$. Considerations similar to those for difference scheme (3.6) show that for

$$\sigma = \frac{1}{2} - \frac{h^2}{12a\tau} \qquad \varphi_i^n = f_i^{n+1/2} + \frac{\tau}{2}\left(\frac{\partial^2 f}{\partial x^2}\right)_i^{n+1/2} \tag{3.10}$$

difference scheme (3.8) is of order $O(\tau^2 + h^2)$. For ease of calculation the right-hand side expression in (3.10) may be written as

$$\varphi_i^n = f_i^{n+1/2} + \frac{1}{12a}(f_{i-1}^{n+1/2} - 2f_i^{n+1/2} + f_{i+1}^{n+1/2})$$

The choice of the right-hand side should be consistent with the approximation-order requirements at the given σ. Thus, in the Crank-Nickolson scheme ($\sigma = 0.5$) it may be assumed that $\varphi_i^n = f_i^{n+1/2}$ or $\varphi_i^n = 0.5(f_i^{n+1} + f_i^n)$.

Stability conditions for difference schemes of the family (3.8) may be obtained by Fourier transformation. The transition factor is then defined by

$$\lambda(\omega) = \frac{1 - (1-\sigma)\zeta(1 - \cos \omega h)}{1 + \sigma\zeta(1 - \cos \omega h)}$$

where $\zeta = 2a\tau/h^2$. Analysis of the expression shows that at $\sigma \geq 0.5$ all the difference schemes of type (3.8) are stable at any τ and h. In particular, this applies to a purely implicit scheme at $\sigma = 1$ and to a symmetric scheme at $\sigma = 0.5$. With $0 \leq \sigma < 0.5$, stability requires the fulfillment of the following inequality:

$$\frac{2a\tau}{h^2} \leq \frac{1}{1-2\sigma} \qquad \sigma \geq \frac{1}{2} - \frac{h^2}{4a\tau} \tag{3.11}$$

The higher-accuracy scheme is stable, since the relation between τ and h conforms with (3.11). Conditional stability is also found at $\sigma < 0$, but such a σ is usually of no practical interest.

All the above conclusions remain valid with nonuniform time steps, i.e., as $\tau^{n+1} = t^{n+1} - t^n$ independent of the step number n. In this case, σ may also be assumed independent of the step number; that is, $\sigma = \sigma^n$. Then stability condition (3.11) is written as

$$\frac{2a\tau^{n+1}}{h^2} \leqslant \frac{1}{1-2\sigma^{n+1}} \qquad \sigma \geqslant \frac{1}{2} - \frac{h^2}{4a\tau^{n+1}} \qquad (3.12)$$

For the higher-accuracy scheme, it should obviously be assumed that

$$\sigma^{n+1} = \frac{1}{2} - \frac{h^2}{12a\tau^{n+1}}$$

It may be of interest to note that if the initial function in a difference scheme of the type (3.8) is positive ($\sigma = 0$), the solution tends to approach the exact solution from below, whereas the classical implicit scheme ($\sigma = 1$) approximates the exact solution from above [11]. Grid methods employing various combinations of these procedures may be expected to give superior results. With this in mind, a symmetrical scheme that holds when $\sigma = 0.5$ seems highly desirable. Higher-accuracy schemes also seem very promising, as they decrease the number of computed grid points and thus decrease the computation time.

Experience shows that difference schemes of the type (3.8) are more advantageous for problems with smooth input data (coefficients, initial conditions). In the case of rapidly changing or discontinuous coefficients and initial functions, multistep difference schemes are to be preferred. A three-step difference scheme suggested by DuFort and Frankel [5] has been widely used for heat-conduction problems. It is of the form

$$\frac{\theta_i^{n+1} - \theta_i^{n-1}}{2\tau} = \frac{a}{h^2}(\theta_{i+1}^n - \theta_i^{n-1} - \theta_i^{n+1} + \theta_{i-1}^n) + f_i^n \qquad (3.13)$$

$$\theta_i^0 = \psi_i \qquad \theta_0^n = \psi_1^n \qquad \theta_J^n = \psi_2^n$$

This is an explicit scheme that is absolutely stable (see Sec. 2.4). The computational algorithm for θ_i^{n+1} is expressed as

$$\theta_i^{n+1} = \frac{2a\tau}{h^2}(\theta_{i+1}^n + \theta_{i-1}^n) + \left(1 - \frac{2a\tau}{h^2}\right)\theta_i^{n-1} + 2\tau f_i^n$$

Calculation of θ_i^{n+1} requires that the temperatures at two previous steps be known. For example, to determine θ_i^2, it is necessary to know both θ_i^0 and θ_i^1. A two-step difference scheme is advisable for θ_i^1. But when θ is changing slowly or when only a steady temperature distribution is desired, one may take $\theta_i^1 = \theta_i^0$.

HEAT-CONDUCTION PROBLEMS

Now let us determine the approximation error of the scheme. To this end, the transformations of the right-hand side of (3.8)

$$\frac{\theta_{i+1}^n - \theta_i^{n-1} - \theta_i^{n+1} + \theta_{i-1}^n}{h^2} = \frac{\theta_{i+1}^n - 2\theta_i^n + \theta_{i-1}^n}{h^2}$$

$$- \frac{\tau^2}{h^2} \frac{\theta_i^{n-1} - 2\theta_i^n + \theta_i^{n+1}}{\tau^2}$$

may be recommended. Now it can easily be noted that the approximation error is defined by

$$R = -a \frac{\tau^2}{h^2} \frac{\partial^2 u}{\partial t^2} + O(\tau^2 + h^2)$$

That is, the scheme has an approximation order of $O(\tau^2 + h^2 + \tau^2/h^2)$. Consequently, the difference equation (3.13) will approximate Eq. (3.4) only when τ vanishes faster than h. If $\tau/h = k = \text{const}$, the difference scheme approximates the hyperbolic equation

$$\frac{\partial \theta}{\partial t} - a \frac{\partial^2 \theta}{\partial x^2} + ak \frac{\partial^2 \theta}{\partial t^2} = f$$

rather than the heat-conduction equation.

Thus, even though difference scheme (3.13) is certainly stable, an accuracy of $O(h^2)$ should ensured by choosing $\tau \approx h^2$. However, for steady-state problems the use of this scheme is quite reasonable.

Implicit three-step weighted schemes may be recommended for the numerical solution of heat-conduction problems with rapidly changing or discontinuous initial data. The computation is generally carried out by one of two types of schemes, symmetrical and nonsymmetrical.

A single-parametric family of symmetrical schemes may be described by

$$\frac{\theta_i^{n+1} - \theta_i^{n-1}}{2\tau} = \frac{\sigma a}{h^2} (\theta_{i+1}^{n+1} - 2\theta_i^{n+1} + \theta_{i-1}^{n+1} + \theta_{i+1}^{n-1} - 2\theta_i^{n-1} + \theta_{i-1}^{n-1})$$

$$+ \frac{(1 - 2\sigma)a}{h^2} (\theta_{i+1}^n - 2\theta_i^n + \theta_{i-1}^n) \quad (3.14)$$

Difference schemes (3.14), with any σ, approximate differential equation (3.4) within the second order of accuracy with respect to τ and h, that is, with an error of $O(\tau^2 + h^2)$. These schemes are stable only for $\sigma > \frac{1}{4}$ [6].

A single-parametric family of nonsymmetrical implicit three-step schemes is of the form

$$\frac{1+\sigma}{\tau}(\theta_i^{n+1} - \theta_i^n) - \frac{\sigma}{\tau}(\theta_i^n - \theta_i^{n-1}) = \frac{a}{h^2}(\theta_{i+1}^{n+1} - 2\theta_i^{n+1} + \theta_{i-1}^{n+1}) + \varphi_i^n \tag{3.15}$$

The order of accuracy of the schemes depends on σ. For $\sigma = 0.5$, their error is of order $O(\tau^2 + h^2)$. In other cases, the error is of order $O(\tau + h^2)$ and stability analysis of the schemes shows that they are stable for $\sigma \geq 0$.

From an error analysis of difference schemes of the form (3.15), it may easily be seen [6] that with

$$\sigma = \frac{1}{2} - \frac{h^2}{12a\tau} \qquad \varphi_i^n = f_i^{n+1} + \frac{h}{12a}\left[\left(\frac{\partial^2 f}{\partial x^2}\right)_i^{n+1} + f_i^{n+1/2}\right]$$

such a scheme has an accuracy of $O(\tau^2 + h^4)$. In all other cases, $\varphi_i^n = f_i^{n+1}$ is advisable.

Schemes of the form (3.14) or (3.15) require that θ_i^{n+1} and θ_i be prescribed at the two previous steps. θ_i^0 is usually prescribed; θ_i^1 may be calculated by the same procedures as for (3.13).

To determine θ_i^{n+1} from (3.14) and (3.15), a system of three-point difference equations of the form

$$A_i \theta_{i-1}^{n+1} + B_i \theta_i^{n+1} + C_i \theta_{i+1}^{n+1} = F(\theta^n, \theta^{n-1})$$

is obtained and solved by the factorization method. Recurrence formulas for the factorization coefficients are of the form

$$\alpha_i = \left(2 + \frac{h^2}{2a\sigma\tau} - \alpha_{i-1}\right)^{-1} \qquad \beta_i = (\beta_{i-1} - F_i)\alpha_i$$

$$F_i = \frac{1-2\sigma}{\sigma}(\theta_{i+1}^n - 2\theta_i^n + \theta_{i-1}^n) + (\theta_{i+1}^{n-1} - 2\theta_i^{n-1} + \theta_{i-1}^{n-1}) \tag{3.16}$$

$$+ \frac{h^2}{2a\sigma\tau}\theta_i^{n-1} + \frac{h^2}{\sigma a}f_i^n$$

for symmetrical schemes; and

#HEAT-CONDUCTION PROBLEMS

$$\alpha_i = \left[2 + \frac{(1+\sigma)h^2}{a\tau} - \alpha_{i-1} \right]^{-1} \qquad \beta_i = (\beta_{i-1} + F_i)\alpha_i$$

$$F_i = \frac{(1+2\sigma)h^2}{a\tau} \theta_i^n - \frac{\sigma h^2}{a\tau} \theta_i^{n-1} + \frac{h^2}{a} f_i^{n+1}$$

(3.17)

for nonsymmetrical schemes. In both cases, the θ_i^{n+1}'s are found from the equation $\theta_i^{n+1} = \alpha_i \theta_{i+1}^{n+1} + \beta_i$. With θ_J^{n+1} known, θ_{J-1}^{n+1}, θ_{J-1}^{n+1}, ..., θ_1^{n+1} are determined in succession.

3.3 CHARACTERISTICS OF NUMERICAL SOLUTIONS OF HEAT-CONDUCTION PROBLEMS IN CYLINDRICAL AND SPHERICAL GEOMETRIES

Let us consider some characteristics of the numerical solutions of heat-conduction problems in cylindrical and spherical geometries. Constant $c\rho$ and κ are again assumed. Here, the heat-propagation processes are governed by the equation

$$\frac{\partial \theta}{\partial t} = \frac{a}{\zeta^m} \frac{\partial}{\partial \zeta} \left(\zeta^m \frac{\partial \theta}{\partial \zeta} \right) + f(\zeta) \qquad 0 \leq \zeta \leq 1 \quad t > 0 \qquad (3.18)$$

which may alternatively be expressed as

$$\frac{\partial \theta}{\partial t} = a \left(\frac{\partial^2 \theta}{\partial \zeta^2} + \frac{m}{\zeta} \frac{\partial \theta}{\partial \zeta} \right) + f(\zeta) \qquad (3.19)$$

where $a = \text{const} > 0$ is the thermal diffusivity, and $f(\zeta)$ is the heat-source distribution function.

With $m = 0$, we obtain Eq. (3.4). For $m = 1$, Eqs. (3.18) and (3.19) describe the temperature distributions in cylindrical geometries with very large lengths compared to diameters. Heat propagation in spherical geometries is governed by these equations where $m = 2$.

It can be seen that when $m = 1$ and $m = 2$, both equations have a singularity at $\zeta = 0$ which must be accounted for in the construction of numerical schemes. To avoid difficulties resulting from the singularity, we ordinarily use the fact that at $\zeta = 0$, the solution is conformable to the conditions $\partial \theta / \partial \zeta = 0$ and $\partial^2 \theta / \partial \zeta^2 = 0$. This allows Eqs. (3.18) and (3.19) to be expressed in a different form. Indeed, because of the two conditions at $\zeta = 0$, the following relation holds:

$$\lim_{\zeta \to 0} \left(\frac{\partial^2 \theta}{\partial \zeta^2} + \frac{m}{\zeta} \frac{\partial \theta}{\partial \zeta} \right) = \frac{\partial^2 \theta}{\partial \zeta^2}\bigg|_{\zeta=0} + m \lim_{\zeta \to 0} \frac{\partial \theta / \partial \zeta}{\zeta} = (1+m) \frac{\partial^2 \theta}{\partial \zeta^2}\bigg|_{\zeta=0}$$

Consequently, at $\zeta = 0$, Eq. (3.18) or (3.19) may be replaced by

$$\frac{\partial \theta}{\partial t} = a(1+m) \frac{\partial^2 \theta}{\partial \zeta^2} + f(0) \tag{3.20}$$

If the difference grid with grid points having coordinates ($\zeta_i = ih$, $t^n = n\tau$), $i = 0, 1, 2, \ldots, J$, $n = 0, 1, 2, \ldots$, is introduced, then for the grid point with index $i = 0$, the following difference relation holds:

$$\frac{\theta_0^{n+1} - \theta_0^n}{\tau} = a(1+m) \frac{\theta_{-1}^n - 2\theta_0^n + \theta_1^n}{h^2} + f(0) \tag{3.21}$$

From (3.21), it immediately follows (using $\theta_{-1} = \theta_1$, because of the symmetrical solution θ) that

$$\theta_0^{n+1} = \frac{2a(1+m)\tau}{h^2} (\theta_1^n - \theta_0^n) - \theta_0^n + \tau f(0) \tag{3.22}$$

The equality

$$\tau \leqslant \frac{h^2}{2a(1+m)} \tag{3.23}$$

is the stability condition for difference equation (3.21).

To weaken the limitation on the parameter τ, the implicit form of Eq. (3.21) may be recommended:

$$\frac{\theta_0^{n+1} - \theta_0^n}{\tau} = a(1+m) \frac{\theta_{-1}^{n+1} - 2\theta_0^{n+1} + \theta_1^{n+1}}{h^2} + f(0) \tag{3.24}$$

In this case θ_0^{n+1} is defined by the formula

$$\theta_0^{n+1} = \frac{\theta_0^n + 2a\tau(1+m)\theta_1^{n+1} + \tau f(0)}{1 + 2a\tau(1+m)h^{-2}} \tag{3.25}$$

HEAT-CONDUCTION PROBLEMS

which implies stability at any τ/h^2. The error of the formula is immediately determined as

$$R = \frac{\tau}{2}\frac{\partial^2 \theta}{\partial t^2} + \frac{a(1+m)h^2}{12}\frac{\partial^4 \theta}{\partial x^4} + O(\tau^2 + h^3)$$

The construction of a difference scheme for Eqs. (3.18) and (3.19), for determining θ_i^{n+1} at other grid points $i = 1, 2, 3, \ldots, J-1$ affords no difficulty. For the numerical solution of the equations, the following two explicit difference schemes are suggested:

$$\frac{\theta_i^{n+1} - \theta_i^n}{\tau} = \frac{a}{\zeta_i^m h^2}\left[\left(\zeta_i + \frac{h}{2}\right)^m (\theta_{i+1}^n - \theta_i^n) - \left(\zeta_i - \frac{h}{2}\right)^m (\theta_i^n - \theta_{i-1}^n)\right] + f_i^n \quad (3.26)$$

and

$$\frac{\theta_i^{n+1} - \theta_i^n}{\tau} = \frac{a}{h^2}(\theta_{i+1}^n - 2\theta_i^n + \theta_{i-1}^n) + \frac{am}{\zeta_i}\frac{\theta_{i+1}^n - \theta_{i-1}^n}{2h} + f_i^n \quad (3.27)$$

With the aid of one of the procedures given in Sec. 2.5, it may be demonstrated that the stability conditions for the equations for $m = 1, 2$ are the same as for the explicit procedure that approximates the heat-conduction equation in Cartesian coordinates ($m = 0$); that is, the schemes are stable when $\tau \leqslant h^2/2a$. Equation (3.26) is an exception. It is stable for $m = 2$ if the more severe condition

$$\tau \leqslant \frac{2h^2}{5a}$$

is satisfied [11].

Examination of the order of the approximation shows that difference equations (3.26) and (3.27) approximate initial equations (3.18) and (3.19) with an error of $O(\tau + h^2 + mh^2/\tau)$. Numerical tests have revealed that difference scheme (3.27) is somewhat more accurate than (3.26).

3.4 APPROXIMATIONS OF BOUNDARY CONDITIONS

The differential heat-conduction equation relates the time and space variations of temperature. It is a mathematical description of the heat transfer within a body. To find the interior temperature field at any time, i.e., to solve a differential problem, it is necessary to know the temperature distribution in the body at an initial time (initial condition), the geometry of the sample, and the law of interaction between the surface and the surrounding fluid (boundary condition).

The initial condition usually refers to a certain time that is assumed as the reference point; the boundary conditions are formulated at the solid surface and are essentially the space boundary conditions.

The initial condition is usually the interior temperature distribution at the initial time $t = 0$:

$$T(x, y, z, 0) = \varphi(x, y, z) \tag{3.28}$$

Boundary conditions may be formulated in different ways:
  1. The surface temperature distribution at any time is prescribed:

$$T(t)|_\Gamma = \psi(t) \tag{3.29}$$

This form of boundary condition is called a boundary condition of the *first kind*. In particular, $T(t) = T_c = \text{const}$ implies that the surface temperature is constant throughout the period of time under consideration.
  2. The heat flux density at every point of the surface is the time function

$$q(t)|_\Gamma = f(t) \tag{3.30}$$

A constant heat flux is the simplest form of this condition.

A modified form of condition (3.30) may be recommended for practical use, namely, heat transfer through the surface is expressed in terms of a temperature gradient along the estimated normal rather than in terms of a heat flux density. Indeed, the basic law of heat conduction states that the heat flux is

$$q = -\lambda \frac{\partial T}{\partial n} \tag{3.31}$$

where λ is the thermal conductivity, and n is a vector of the external normal. Therefore, (3.30) may be replaced by

$$\left.\frac{\partial T}{\partial n}\right|_\Gamma = \bar{f}(t) \tag{3.32}$$

HEAT-CONDUCTION PROBLEMS

where $\bar{f}(t) = -\lambda f(t)$. In particular, the condition $\partial T/\partial n = 0$ implies no heat transfer (the solid surface is thermally insulated).

Expressions (3.30) and (3.32) are called boundary conditions of the *second kind*.

3. The *third-kind* boundary condition is the law of convective heat transfer between the solid surface and the ambient fluid at a constant heat flux. Here, the heat quantity transferred per unit time from a unit surface area to the ambient fluid with temperature T_c during cooling ($T_\Gamma > T_c$) is proportional to the temperature difference between the solid surface and the surrounding medium: $q_\Gamma = \alpha(T_\Gamma - T_c)$, where α is the heat-transfer coefficient. The heat-transfer coefficient α depends on thermophysical properties of the sample and the ambient fluid, and on the temperature. Naturally, for heating ($T_\Gamma < T_c$), a similar expression holds in which T_Γ and T_c are interchanged. With these relations, condition (3.31) may be written as

$$\left(\alpha \frac{\partial T}{\partial n} + \alpha T\right)_\Gamma = \alpha T_c \qquad (3.33)$$

In some cases, boundary conditions of the third kind describe temperature behavior with radiant heat transfer. According to the Stefan-Boltzmann law, the radiant heat flux between two surfaces is

$$q(t)|_\Gamma = \sigma^*(T_\Gamma^4 - T_c) \qquad (3.34)$$

where σ^* is the dimensionless radiation coefficient, and T_c is the absolute temperature of the heat receiver. For a small temperature difference $T_\Gamma - T_c$, relation (3.34) may be approximated as

$$q_\Gamma = \sigma^*(T_\Gamma^2 + T_c^2)(T_\Gamma + T_c)(T_\Gamma - T_c) = \alpha(T)(T_\Gamma - T_c) \qquad (3.35)$$

where $\alpha(T)$ is the radiant heat-transfer coefficient.

4. Finally, boundary conditions of the *fourth kind* describe heat transfer with the surrounding medium or heat transfer between two contacting solids with equal temperatures at their surfaces. In this case, the boundary conditions prescribe equal temperatures and heat fluxes from both sides of the boundary:

$$T_1(t)_\Gamma = T_2(t)_\Gamma \qquad -\lambda_1 \left(\frac{\partial T_1}{\partial n}\right)_\Gamma = -\lambda_2 \left(\frac{\partial T_2}{\partial n}\right)_\Gamma \qquad (3.36)$$

In the numerical solution of heat-transfer problems, a differential form is replaced with a difference algorithm. Along with the substitution of a difference equation for a differential equation, it is necessary to approximate the boundary conditions with finite-difference relations. The accuracy of the resultant numerical solution depends on both the error in approximating the original differential equation with the difference form and the error resulting from approximation of the boundary conditions. For example, if a differential equation is approximated with an error of order $O(h^2)$, and the boundary conditions with an accuracy $O(h)$, the order of accuracy of the resultant numerical solution is not higher than $O(h)$.

Let us consider some methods for increasing the order of approximation of boundary conditions, for given grid points. The approximation of first-kind boundary conditions obviously affords no mathematical difficulties. Indeed, if the boundary of the grid region is the same as the real boundary of the body inside which the temperature changes are to be determined, the boundary conditions are approximated exactly. The functions of the exact solution at the points of the real boundary corresponding to the chosen grid points are usually taken as boundary grid functions. When the boundary of the grid region is not the same as the real boundary, the boundary conditions for the difference problem are obtained by interpolation. In this case, the accuracy of the difference boundary conditions depends on the accuracy of interpolation.

Let us consider some approximation methods applicable to boundary conditions of the third kind. The discussion that follows is also valid for boundary conditions of the second kind, since they are special cases of third-kind boundary conditions. For a proper choice of approximation method for the boundary conditions, the characteristics of the particular problem and of the finite-difference approximation to the original differential equations must be taken into account.

As an example, suppose an approximate solution of the heat-conduction equation

$$\frac{\partial \theta}{\partial t} = \frac{\partial^2 \theta}{\partial x^2} + f(x, t) \quad 0 \leqslant x \leqslant 1 \quad 0 \leqslant t \leqslant T \tag{3.37}$$

with the boundary conditions

$$\frac{\partial \theta(0, t)}{\partial x} = b_1 \theta(0, t) + \psi_1(t) \quad \theta(1, t) = \psi_2(t) \tag{3.38}$$

HEAT-CONDUCTION PROBLEMS

is to be found. A regular grid whose grid points have the coordinates $(x_i = ih,\ t^n_* = n\tau)$, $i = 0, 1, 2, \ldots, J$, $n = 0, 1, 2, \ldots$ is adopted, and an explicit two-step scheme of order $O(\tau + h^2)$ is constructed.

A difference approximation of the same order of accuracy is constructed for the boundary condition at $x = 0$. The approximation of the derivative with respect to x in the boundary condition poses some difficulty. An approximation of the second order with respect to h on three points of the space grid is certainly not very difficult. However, for large temperature gradients in the boundary region, the approximation will produce a large absolute value of error. In such cases, the boundary conditions should be approximated on a pattern with as few grid points as possible.

The approximation method suggested for boundary conditions of the second and third kinds is based on the use of the initial differential equation to obtain a more accurate difference approximation of the conditions, of the first order of accuracy.

Consider the difference relation that approximates the derivative $\partial\theta/\partial x$ at $x = 0$ with $O(h)$, and let us analyze the main terms of the error

$$\frac{\theta_1 - \theta_0}{h} = \frac{\partial\theta(0, t)}{\partial x} + \frac{h}{2}\frac{\partial^2\theta(0, t)}{\partial x^2} + O(h^2) \tag{3.39}$$

Using (3.37), we may write

$$\frac{\partial^2\theta(0, t)}{\partial x^2} = \frac{\partial\theta(0, t)}{\partial t} - f(0, t)$$

Substitution of this expression into (3.39) gives

$$\frac{\theta_1 - \theta_2}{h} - \frac{h}{2}\left[\frac{\partial\theta(0, t)}{\partial t} - f(0, t)\right] = \frac{\partial\theta(0, t)}{\partial x} + O(h^2)$$

The expression on the left-hand side approximates the derivative $\partial\theta/\partial x$ at $x = 0$ with error $O(h^2)$. Substitution of this expression, with the derivative $\partial\theta(0, t)/\partial t$ replaced by the differencing relation $(\theta_0^{n+1} - \theta_0^n)/\tau$, into (3.38) gives the differencing relation for the boundary condition at $x = 0$ with error $O(h^2)$. Since the system (3.37)–(3.38) is calculated with an explicit procedure, the boundary conditions should also be expressed in an explicit form, as

$$\frac{\theta_1{}^n - \theta_0{}^n}{h} - \frac{h}{2}\left(\frac{\theta_0{}^{n+1} - \theta_0{}^n}{\tau} - f_0{}^n\right) = \mathcal{b}_1\theta_0{}^n + \psi_1{}^n \qquad (3.40)$$

or, alternatively, as

$$\theta_0{}^{n+1} = \zeta\theta_1{}^n + [1 - \zeta(1 + \mathcal{b}_1 h)]\theta_0{}^n - \tau f_0{}^n + \zeta h \psi_1{}^n$$

where $\zeta = 2\tau/h^2$

The boundary condition expression (3.40) also has the order of approximation $O(\tau + h^2)$.

When an implicit procedure is used in place of (3.40),

$$\frac{\theta_1{}^{n+1} - \theta_0{}^{n+1}}{h} - \frac{h}{2}\left(\frac{\theta_0{}^{n+1} - \theta_0{}^n}{\tau} - f_0{}^{n+1}\right) = \mathcal{b}_1\theta_0{}^{n+1} + \psi_1{}^{n+1} \qquad (3.41)$$

should be considered. Simple manipulations give the condition

$$\theta_0{}^{n+1} = \frac{\zeta}{1 + \zeta(1 + h\mathcal{b}_1)} \theta_1{}^{n+1} + \frac{\theta_0{}^n + \zeta h \psi_1{}^{n+1} - \tau f_0{}^{n+1}}{1 + \zeta(1 + h\mathcal{b}_1)}$$

where again $\zeta = 2\tau/h^2$. Condition (3.41) also has an error of $O(\tau + h^2)$. If the difference equations are solved by the factorization method, then obviously

$$\alpha_0 = \frac{\zeta}{1 + \zeta(1 + h\mathcal{b}_1)} \qquad \beta_0 = \frac{\alpha_0}{\zeta}(\theta_0{}^n - \zeta h \psi_1{}^{n+1} - \tau f_0{}^{n+1}) \qquad (3.42)$$

If the weighted difference scheme (3.8) is used to solve (3.37)–(3.38), then it may similarly be demonstrated that the difference condition

$$\sigma\left(\frac{\theta_1{}^{n+1} - \theta_0{}^{n+1}}{h} - \mathcal{b}_1\theta_0{}^{n+1}\right) + (1 - \sigma)\frac{\theta_1{}^n - \theta_0{}^n}{h} - \mathcal{b}_1\theta_0{}^n$$

$$= \frac{h}{2}\left(\frac{\theta_0{}^{n+1} - \theta_0{}^n}{\tau} - f_0{}^{n+1/2}\right) + \psi_1{}^{n+1/2} \qquad (3.43)$$

where $f_0{}^{n+1/2} = f(0, t^{n+1/2})$
$\psi_1{}^{n+1/2} = \psi_1(t^{n+1/2})$

approximates boundary condition (3.38) at $x = 0$ with the same order as system (3.8) approximates Eq. (3.37) for fixed σ.

HEAT-CONDUCTION PROBLEMS

Condition (3.43) will now be reduced to a form convenient for computation. It will be solved with respect to θ_0^{n+1}:

$$\theta_0^{n+1} = \alpha_0 \theta_1^{n+1} + \beta_0$$

$$\alpha_0 = \sigma \left[\sigma(1 + h\boldsymbol{6}_1) + \frac{h^2}{2\tau} \right]^{-1}$$

$$\beta_0 = \left\{ (1-\sigma)\theta_1^n + \left[\frac{h^2}{2\tau} - (1-\sigma)(1 + h\boldsymbol{6}_1) \right] \theta_0^n \right.$$

$$\left. - 0.5 h^2 f_0^{n+1/2} + h\psi_1^{n+1/2} \right\} \frac{\alpha_0}{\sigma}$$

It may be easily seen that for $\sigma = 0$, condition (3.40) is obtained; and for $\sigma = 1$ condition (3.41) follows.

In problem (3.37)–(3.38) the first-kind boundary condition is given at the boundary $x = 1$. If the third-kind boundary condition is given at $x = 1$, it may be approximated in exactly the same way as the condition at $x = 0$. For example, if at $x = 1$ the boundary condition is of the form

$$-\frac{\partial \theta(1, t)}{\partial x} = \boldsymbol{6}_2 \theta(1, t) + \psi_2(t) \tag{3.44}$$

this may be approximated by the expression

$$\theta_J^{n+1} = \bar{\alpha}_0 \theta(1, t) + \psi_2(t) \tag{3.45}$$

where

$$\bar{\alpha}_0 = \sigma \left[\sigma(1 + h\boldsymbol{6}_2) + \frac{h^2}{2\tau} \right]^{-1}$$

$$\bar{\beta}_0 = \left\{ (1-\sigma)\theta_{J-1}^n + \left[\frac{h^2}{2\tau} - (1-\sigma)(1 + h\boldsymbol{6}_2) \right] \theta_J^n \right.$$

$$\left. - 0.5 h^2 f_J^{n+1/2} + h\psi_2^{n+1/2} \right\} \frac{\bar{\alpha}_0}{\sigma}$$

The approximation error involved with difference boundary condition (3.45) depends on the parameter σ. There is complete similarity with condition (3.43).

3.5 FINITE-DIFFERENCE PROCEDURES FOR HEAT-CONDUCTION EQUATIONS WITH VARIABLE COEFFICIENTS

In previous sections, finite-difference procedures were considered for heat-conduction problems with constant heat capacities, densities, and thermal conductivities. This situation is very seldom encountered in reality. Instead, the above quantities depend both on space variables and time, and we must deal with heat-conduction equations with variable coefficients.

In the general case, especially when a solution is sought for a wide temperature range, the thermophysical properties of the medium (such as heat capacity and thermal conductivity) also depend on temperature, so that the problem is nonlinear. Nonlinear heat-conduction problems will be discussed in later sections. Here, some useful and relatively simple difference methods for one-dimensional linear problems with variable coefficients will be discussed.

Let us consider such a problem. The solution of the equation

$$\frac{\partial \theta}{\partial t} = \frac{\partial}{\partial x}\left[\kappa(x, t) \frac{\partial \theta}{\partial x}\right] + f(x, t) \tag{3.46}$$

which is continuous in the region $0 \leqslant x \leqslant 1$, $0 \leqslant t \leqslant T$, is to be found subject to the boundary conditions

$$\theta(x, 0) = \psi(x) \qquad \theta(0, t) = \psi_1(t) \qquad \theta(1, t) = \psi_2(t) \tag{3.47}$$

The heat-conduction coefficient is assumed to be the bounded function

$$0 \leqslant c_1 \leqslant \kappa(x, t) \leqslant c_2 \tag{3.48}$$

where c_1 and c_2 are constant positive values.

For the numerical solution, a family of conservative finite-difference schemes will be built up with the aid of the integrointerpolation procedure described in Sec. 2.6. A difference grid with grid points having coordinates (x_i, t^n), $i = 0, 1, 2, \ldots, J$, $n = 0, 1, 2, \ldots$, will be used. A time $t = \bar{t} = t^{n+1/2}$ is fixed, and on the basis of Eq. (3.46) we make a balance for an element $l_i(x_{i-1/2} \leqslant x \leqslant x_{i+1/2})$:

$$\int_{l_i} \frac{\partial \theta(x, \bar{t})}{\partial t} dx = q_{i+1/2} - q_{i-1/2} + \int_{l_i} f(x, \bar{t}) dx \tag{3.49}$$

HEAT-CONDUCTION PROBLEMS

where $q_{i\pm 1/2} = \left[\kappa(x, \bar{t}) \dfrac{\partial \theta(x, \bar{t})}{\partial x} \right]_{x=x_{i\pm 1/2}}$

Approximation of the terms in (3.49) gives

$$\int_{l_i} \frac{\partial \theta(x, \bar{t})}{\partial t} dx \approx h \frac{\theta_i^{n+1} - \theta_i^n}{\tau}$$

$$\int_{l_i} f(x, \bar{t}) dx \approx h \varphi_i^n$$

$$q_{i-1/2} \approx a_i \left[\sigma \frac{\theta_i^{n+1} - \theta_{i-1}^{n+1}}{h} + (1 - \sigma) \frac{\theta_i^n - \theta_{i-1}^n}{h} \right]$$

$$q_{i+1/2} \approx a_{i+1} \left[\sigma \frac{\theta_{i+1}^{n+1} - \theta_i^{n+1}}{h} + (1 - \sigma) \frac{\theta_{i+1}^n - \theta_i^n}{h} \right]$$

where σ is a material parameter. Coefficients a_i and a_{i+1} are expressed in terms of the function $\kappa(x, t)$ in such a way that the approximation is of the order $O(h^2)$ at any t. This will be further discussed a bit later.

Substitution of the resultant expressions into Eq. (3.49) yields the following homogeneous conservative difference scheme:

$$\frac{\theta_i^{n+1} - \theta_i^n}{\tau} = \frac{1}{h} \left\{ a_{i+1} \left[\sigma \frac{\theta_{i+1}^{n+1} - \theta_i^{n+1}}{h} + (1 - \sigma) \frac{\theta_{i+1}^n - \theta_i^n}{h} \right] \right.$$

$$\left. - a_i \left[\sigma \frac{\theta_i^{n+1} - \theta_{i-1}^{n+1}}{h} + (1 - \sigma) \frac{\theta_i^n - \theta_{i-1}^n}{h} \right] \right\} + \varphi_i^n \quad (3.50)$$

$$\theta_i^0 = \psi_i \qquad \theta_0^n = \psi_1^n \qquad \theta_J^n = \psi_2^n$$

The initial and boundary conditions are approximated on the grid exactly. When scheme (3.50) is utilized in practice, the following expressions for a_i, a_{i+1}, and φ_i^n are recommended [6]:

$$a_i = \kappa(x_{i-1/2}, \bar{t}) \qquad\qquad a_{i+1} = \kappa(x_{i+1/2}, \bar{t}) \qquad (3.51a)$$

$$a_i = \frac{1}{2} [\kappa(x_i, \bar{t}) + \kappa(x_{i-1}, \bar{t})] \qquad a_{i+1} = \frac{1}{2} [\kappa(x_{i+1}, \bar{t}) + \kappa(x_i, \bar{t})]$$

$$(3.51b)$$

$$a_i = 2\left[\frac{1}{\kappa(x_i,\bar{t})} + \frac{1}{\kappa(x_{i-1},\bar{t})}\right]^{-1} \quad a_{i+1} = 2\left[\frac{1}{\kappa(x_i,\bar{t})} + \frac{1}{\kappa(x_{i+1},\bar{t})}\right]^{-1}$$
(3.51c)

$$\varphi_i^n = f(x_i, \bar{t}) = f_i^{n+1/2}$$

With $\kappa(x, t) = \text{const}$, difference scheme (3.50) becomes the weighted scheme for a heat-conduction equation with constant coefficients, which was treated in the previous sections.

The above difference scheme may be rearranged to a form convenient for computation:

$$-F_i^n = a_i \theta_{i-1}^{n+1} - \left(a_i + a_{i+1} + \frac{h^2}{\sigma\tau}\right)\theta_i^{n+1} + a_{i+1}\theta_{i+1}^{n+1}$$

$$F_i^n = \frac{h}{\sigma\tau}\theta_i^n + \frac{1-\sigma}{\sigma}[a_i\theta_{i-1}^n - (a_i + a_{i+1})\theta_i^n + a_{i+1}\theta_{i+1}^n] + \frac{h^2}{\sigma}\varphi_i^n$$
(3.52)

A system of difference equations with boundary conditions $\theta_0^{n+1} = \psi_1^{n+1}$ and $\theta_J^{n+1} = \psi_2^{n+1}$ is obtained for θ_i^{n+1} at the $(n+1)$th time step. The system may be solved by the factorization method. For $\sigma > 0$, the stability conditions for the method (see Sec. 2.7) are satisfied. The recurrence formulas for the factorization coefficients are

$$\alpha_i = a_{i+1}\left[\frac{h^2}{\sigma\tau} + a_{i+1} + a_i(1 - \alpha_{i-1})\right]^{-1} \quad \beta_i = (a_i\beta_{i-1} + F_i^n)\frac{\alpha_i}{a_{i+1}} \quad (3.53)$$

From the boundary conditions at $x = 0$, we have $\alpha_0 = 0$ and $\beta_0 = \psi_1^n$. Then $\alpha_1, \beta_1, \alpha_2, \beta_2, \ldots, \alpha_{J-1}, \beta_{J-1}$ are successively determined from (3.53). Next, since $\theta_J^{n+1} = \psi_2^{n+1}$ (the second boundary condition), $\theta_{J-1}^{n+1}, \theta_{J-2}^{n+1}, \ldots, \theta_1^{n+1}$ are successively found with the formula

$$\theta_{i-1}^{n+1} = \alpha_{i-1}\theta_i^{n+1} + \beta_{i-1}$$

For the calculation of F_i^n on the right-hand side, the recurrence formula

$$F_i^n = \frac{h^2}{\sigma^2\tau}\theta_i^n - \frac{1-\sigma}{\sigma}F_i^{n-1} + \frac{h^2}{\sigma}\varphi_i^n \quad (3.54)$$

may sometimes be useful.

HEAT-CONDUCTION PROBLEMS

It should, however, be remembered that for this formula to be used, the block of values F_i^{n-1} consisting of numbers $J+1$ must be stored. When the computer storage admits this, the use of formula (3.54) saves a considerable amount of time.

If a_i and φ_i approximate κ_i and f_i, respectively, with error $O(h^2)$, then difference scheme (3.50) may be shown to approximate differential problem (3.46)-(3.47) with error $O(\tau + h^2)$ for $\sigma \neq 0.5$ and with error $O(\tau^2 + h^2)$ for $\sigma = 0.5$.

The case of $\sigma = 0$ leads to an explicit scheme. Stability analysis reveals that it is stable at

$$\tau \leqslant \frac{h^2}{2c_2} \tag{3.55}$$

The admissible time step τ strongly depends on the coefficient $\kappa(x, t)$. The use of explicit schemes for heat-conduction problems with large thermal diffusivities does not seem advisable.

For $\sigma \neq 0$, formula (3.50) is an implicit scheme. It may be demonstrated [6] that a condition of sufficiency for these schemes is the inequality

$$\sigma \geqslant \frac{1}{2} - \frac{h^2}{4\tau c_2} \tag{3.56}$$

Thus, for $\sigma \geqslant 0{,}5$, difference scheme (3.50) is always stable. For the scheme to be stable for $\sigma < 0.5$, the time step τ should be chosen so that the following relation holds:

$$\tau \leqslant \frac{h^2}{2(1 - 2\sigma)c_2} \tag{3.57}$$

In engineering practice it is often of interest to calculate the heat transfer in regions consisting of two or more inhomogeneous media with different thermophysical properties, particularly different thermal diffusivities. In such a case, the coefficient $\kappa(x, t)$ in Eq. (3.46) undergoes a discontinuity of the first kind at the interface; for example, if x_i is the discontinuity point, then $\kappa(x_i - 0, t) \neq \kappa(x_i + 0, t)$. At discontinuity points, the following conjugate conditions must be observed:

$$\theta(x_i - 0, t) = \theta(x_i + 0, t)$$
$$\kappa(x_i - 0, t) \frac{\partial \theta(x_i - 0, t)}{\partial x} = \kappa(x_i + 0, t) \frac{\partial \theta(x_i + 0, t)}{\partial x} \qquad (3.58)$$

Now let us return to difference scheme (3.50). A unique advantage of the scheme is that this equation may also be used in the case when $\kappa(x, t)$ and $f(x, t)$ have discontinuities of the first kind on the straight line $x = \text{const}$. In this case the grid should be constructed so that the grid points coincide with the discontinuity points. For the function φ_i^n at the discontinuity points of Eq. (3.50), one should take the arithmetic mean of its extreme values:

$$\varphi_i^n = 0.5[f(x_i - 0, t^{n+1/2}) + f(x_i + 0, t^{n+1/2})]$$

In addition, the coefficients a_i should be chosen from condition (3.58), e.g.,

$$a_i = \kappa(x_{i-1/2}, t^{n+1/2}) \qquad a_{i+1} = \kappa(x_{i+1/2}, t^{n+1/2})$$

With discontinuous coefficients, the accuracy of difference schemes of the type of (3.50) is somewhat lower, but the approximation order in all cases will remain at least at $O(\tau + h)$.

Grids with nonuniform t and x are frequently applied in practice. The nonuniformity of t does not change the above formulas and estimates. One need only remember that step $\tau = \tau^n$ depends on the step number n, whereas the approximation order for the time derivative remains the same.

Let us now consider a grid involving a nonuniform space derivative. We choose a grid x_i, $i = 0, 1, 2, \ldots, J$, with steps $h_i = x_i - x_{i-1}$ for the interval $0 \leq x \leq 1$.

In the same way as difference scheme (3.50) was constructed for problem (3.46)–(3.47), a difference scheme of the following form is constructed:

$$\frac{\theta_i^{n+1} - \theta_i^n}{\tau} = \frac{1}{\hbar_i} \left\{ a_{i+1} \left[\sigma \frac{\theta_{i+1}^{n+1} - \theta_i^{n+1}}{h} + (1-\sigma) \frac{\theta_{i+1}^n - \theta_i^n}{h} \right] \right.$$
$$\left. - a_i \left[\sigma \frac{\theta_i^{n+1} - \theta_{i-1}^{n+1}}{h} + (1-\sigma) \frac{\theta_i^n - \theta_{i-1}^n}{h} \right] \right\} + \varphi_i^n \qquad (3.59)$$

$$\theta_i^0 = \psi_i \qquad \theta_0^n = \psi_1^n \qquad \theta_J^n = \psi_2^n$$

where $\hbar = 0.5(h_i + h_{i+1})$

HEAT-CONDUCTION PROBLEMS

Now, very simple and widely used formulas for the coefficients a_i will be presented. It must be remembered that the function $\kappa(x, t)$ may have a finite number of discontinuities of the first kind, and at the discontinuity points (if any) the conjugate conditions (3.58) are satisfied. The formulas are

$$a_i = 0.5[\kappa(x_i - 0, \bar{t}) + \kappa(x_i + 0, \bar{t})]$$

$$a_i = \left(\frac{1}{h} \int_{x_{i-1}}^{x_i} \frac{dx}{\kappa(x, \bar{t})}\right)^{-1} \tag{3.60}$$

$$a_i = \frac{2\kappa(x_i - 0, \bar{t})\kappa(x_i + 0, \bar{t})}{\kappa(x_i - 0, \bar{t}) + \kappa(x_i + 0, \bar{t})}$$

The following expression is generally used for φ_i^n:

$$\varphi_i^n = \frac{1}{2\hbar_i}[h_i f(x_i - 0, \bar{t}) + h_{i+1} f(x_i + 0, \bar{t})] \tag{3.61}$$

With θ_i^n known at time t^n, the θ_i^{n+1} at the $(n+1)$th time step are evaluated by solving the system of difference equations

$$-F_i^n = \frac{a_i}{h_i}\theta_{i-1}^{n+1} - \left(\frac{a_i}{h_i} + \frac{a_{i+1}}{h_{i+1}} + \frac{\hbar}{\sigma\tau}\right)\theta_i^{n+1} + \frac{a_{i+1}}{h_{i+1}}\theta_{i+1}^{n+1}$$

$$F_i^n = \left[\frac{\hbar}{\sigma\tau} - \frac{1-\sigma}{\sigma}\left(\frac{a_i}{h_i} + \frac{a_{i+1}}{h_{i+1}}\right)\right]\theta_i^n + \frac{1-\sigma}{\sigma}\left(\frac{a_i}{h_i}\theta_{i-1}^n + \frac{a_{i+1}}{h_{i+1}}\theta_{i+1}^n\right) + \varphi_i^n \tag{3.62}$$

by the factorization method. Factorization coefficients are evaluated from the formulas

$$\alpha_i = \frac{a_{i+1}}{h_{i+1}}\left[\frac{\hbar_i}{\sigma\tau} + \frac{a_{i+1}}{h_{i+1}} + \frac{a_i}{h_i}(1 - \alpha_{i-1})\right]^{-1} \qquad \beta_i = \left(\frac{a_i}{h_i}\beta_{i-1} + F_i^n\right)\frac{h_{i+1}}{a_{i+1}}\alpha_i$$

If the coefficients $\kappa(x, t)$ and $f(x, t)$ are smooth, and a and φ are approximated with the order $O(\tau + h^2)$, then difference scheme (3.59), with any set of nonuniform grids, uniformly converges at the rate $O(\tau + h^2)$, $h = \max_i h_i$, with

$$\sigma \geqslant \frac{1}{2} - \frac{h^2}{4\tau c_2} \tag{3.63}$$

For $\sigma = 0.5$, the rate of convergence is $O(\tau^2 + h^2)$.

If the functions $\kappa(x, t)$ and $f(x, t)$ have a finite number of first-kind discontinuities on straight lines $x = $ const, if the conjugate conditions (3.58) are satisfied at the discontinuity points, and if the discontinuity lines of the functions $\kappa(x, t)$ and $f(x, t)$ pass through the difference grid points, then the rate of convergence of difference scheme (3.59) is not less than $O(\tau + h)$, provided that condition (3.63) is satisfied and a_i and φ_i are reasonably chosen.

To conclude this section, it should be mentioned that there are a number of numerical schemes for solving problems of the type of (3.46)–(3.47). Among the advantages of the methods just discussed is the fact that, because of their conservative nature, they give the heat-flux balance for the problem quite accurately. Moreover, they possess a rather high accuracy and remain stable for relatively large time steps. The value $\sigma = 0.5$ may be recommended for most problems.

3.6 FRACTIONAL-STEP METHOD

Real physical processes involve time and three-dimensional space. The numerical analysis of real physical problems should therefore be based on three-dimensional mathematical models. However, in such cases, additional difficulties because of the three-dimensionality and stability are normally encountered. An increase in the number of dimensions requires that equations have three or more independent variables, compared with two variables in the one-dimensional case. Hence, a differential problem leads to a system of difference equations with approximately h^{-p} unknowns (where p is the number of dimensions). Care should therefore be exercised in choosing a difference scheme, since the number of operations involved may be beyond reasonable limits.

The numerical stability of the difference scheme is also a problem in the case of two or more dimensions. This, however, applies only to very simple computational procedures. Thus, for example, if the stability of an explicit scheme in a one-dimensional case is governed by the condition $\tau < h^2/2$, then, in a two-dimensional problem this condition becomes $\tau < h^2/4$.

HEAT-CONDUCTION PROBLEMS

Considerations of economy are very important to a correct choice of difference schemes for multidimensional problems. Economy here is understood to mean the number of arithmetic operations necessary for one time (iteration) step.

Let us consider a simple two-dimensional heat-conduction problem

$$\frac{\partial \theta}{\partial t} = \frac{\partial^2 \theta}{\partial x^2} + \frac{\partial^2 \theta}{\partial y^2} \quad 0 \leq t \leq T \quad -\infty < x, y < \infty \tag{3.64}$$

$$\theta(x, y, 0) = \psi(x, y)$$

A difference grid with finite steps Δx, Δy, τ is used to construct the derivatives at x, y, t, respectively. For convenience, the following notations are used:

$$\theta_{i,j}^n = \theta(i\,\Delta x, j\,\Delta y, n\tau)$$

$$\delta_x \theta = \frac{\theta_{i+1,j} - 2\theta_{i,j} + \theta_{i-1,j}}{(\Delta x)^2}$$

$$\delta_y \theta = \frac{\theta_{i,j+1} - 2\theta_{i,j} + \theta_{i,j-1}}{(\Delta y)^2}$$

The finite steps for the space variables x and y are assumed equal; that is, $\Delta x = \Delta y = h$.

The construction of a difference scheme for problem (3.64) entails no special difficulties. Most of the difference procedures discussed in previous sections may be easily extended to the case of two and more dimensions. It should be remembered, however, that in this case they may not be so effective as for one-dimensional problems. This will be illustrated with a simple example. In a manner analogous to the one-dimensional case, the two schemes for problem (3.64) can be written as

$$\frac{\theta_{i,j}^{n+1} - \theta_{i,j}^n}{\tau} = \delta_x \theta^n + \delta_y \theta^n \tag{3.65}$$

$$\frac{\theta_{i,j}^{n+1} - \theta_{i,j}^n}{\tau} = \delta_x \theta^{n+1} + \delta_y \theta^{n+1} \tag{3.66}$$

The first is an explicit scheme. The calculation of $\theta_{i,j}^{n+1}$, with $\theta_{i,j}^n$ known, requires a number of arithmetic operations proportional to the number of unknown quantities ($\sim h^{-2}$). In this respect, this explicit scheme cannot be optimal, and it is stable only with very severe limitations on the time steps ($\tau < h^2/4$). Therefore, the application of explicit procedures to two-dimensional and multidimensional problems may often be unreasonable.

The second scheme is implicit. It is stable for any τ and h. The calculation of $\theta_{i,j}^{n+1}$ requires the solution of approximately h^{-2} equations. Whereas for a one-dimensional case the use of the scalar factorization method was possible, for two or more dimensions this method is totally inapplicable. Other methods for going from $\theta_{i,j}^n$ to $\theta_{i,j}^{n+1}$ require a number of arithmetic operations of order h^{-6}. Thus, the use of scheme (3.66) leads to a tremendous number of computations, so it is quite inefficient.

Recently, difference procedures based on fractional steps have been successfully used [23] to solve multidimensional problems. The procedures combine the advantages of explicit and implicit schemes: They are certainly stable, and the transition from one step to the next requires a number of arithmetic operations proportional to the number of grid points ($\sim h^{-2}$). The economy of difference procedures based on the fractional-step method results from the fact that multidimensional problems can be reduced to one-dimensional cases, with the latter being solved by the factorization method.

Since such methods appear, at present, to be most efficient for solving multidimensional problems, one fractional-step method will be described in detail. It is known as the *splitting-step* method.

The basic idea of this method may be described as follows: The solution of problem (3.64) is to be found on the time interval $t^n \leqslant t \leqslant t^{n+1}$. The $\theta(x, y, t^n)$ are assumed known, and the $\theta(x, y, t^{n+1})$'s are to be found. In place of (3.64), we consider two problems:

$$\frac{\partial u}{\partial t} = \frac{\partial^2 u}{\partial x^2} \quad t^n \leqslant t \leqslant t^{n+1} \quad u(x, y, t^n) = \theta(x, y, t^n) \tag{3.67}$$

$$\frac{\partial v}{\partial t} = \frac{\partial^2 v}{\partial y^2} \quad t^n \leqslant t \leqslant t^{n+1} \quad v(x, y, t^n) = u(x, y, t^{n+1}) \tag{3.68}$$

A Taylor-series expansion readily shows that if $\theta(x, y, t)$ is the solution of problem (3.64),

$$v(x, y, t^{n+1}) = \theta(x, y, t^{n+1}) + O(h^2)$$

HEAT-CONDUCTION PROBLEMS

For each time interval $t^n \leq t \leq t^{n+1}$, this equation allows the successive solution of problems (3.67) and (3.68) instead of (3.64).

This splitting of the two-dimensional problem (3.64) into two one-dimensional problems may be considered as an approximate representation of two heat-propagation processes over the xy plane during a time $t^n \leq t \leq t^{n+1}$. In the first process, described by Eq. (3.67), fictitious diaphragms that do not let heat propagate along the y axis are introduced. Then, after time τ passed, other fictitious diaphragms that do not let heat propagate along the x axis are introduced, and the other diaphragms are removed. This processes is described by Eq. (3.68).

Problem (3.64) may be split in several different ways. The split presented here is based on physical considerations and was chosen for illustrative purposes.

The solution of Eqs. (3.67) and (3.68) may be approximated with difference equations. A difference splitting scheme is then obtained, from which the $\theta_{i,j}^{n+1}$'s are calculated in two stages, with $\theta_{i,j}^n$ known. In the first stage, the $u_{i,j}^n = \theta_{i,j}^n$'s are calculated from given $u_{i,j}^{n+1}$; in the second stage, the $v_{i,j}^n = u_{i,j}^{n+1}$'s are determined with known $v_{i,j}^{n+1} \approx \theta_{i,j}^{n+1}$.

The above assumptions are heuristic. After the difference splitting scheme for (3.64) is constructed, its stability and approximations should be tested. A number of finite-difference schemes for heat-conduction equations make use of the fractional-step method. Some of them will be discussed below.

3.7 TWO-DIMENSIONAL HEAT-CONDUCTION PROBLEMS

Real heat-transfer processes take place in a three-dimensional space that has a certain effect on the process. Three-dimensional effects should be accounted for as much as possible in the analysis of heat-conduction problems, and the problems must be formulated as three-dimensional. However, the solution of three-dimensional problems requires a tremendous amount of computational work.

In practice, cases may be encountered in which quantities change minimally with respect to one of the dimensions. Then, the three-dimensional problem may be reduced to a two-dimensional problem whose solution is much simpler. For example, suppose we wish to analyze heat propagation in a very thin cylinder with a uniform cross section and a length much larger than the diameter. It is clear from physical considerations that the temperature distributions in the cross sections of the cylinder far from its faces are almost all the same. Therefore, we may consider the temperature distribution in one of these cross sections only.

Next we shall discuss the methods that may be used for the approximate solution of two-dimensional heat-conduction problems on high-speed computers. Schemes based on the fractional-step method are most efficient. A large number of such schemes are available in the literature. The most widely used modifications suitable for heat-conduction problems will be considered here.

The solution $\theta(x, y, t)$ of a two-dimensional heat-conduction problem inside the rectangular region $0 \leqslant x \leqslant l_x$, $0 \leqslant y \leqslant l_y$ may be found from

$$\frac{\partial \theta}{\partial t} = \frac{\partial}{\partial x}\left[\kappa_1(x, y, t) \frac{\partial \theta}{\partial x}\right] + \frac{\partial}{\partial y}\left[\kappa_2(x, y, t) \frac{\partial \theta}{\partial y}\right] + f \tag{3.69}$$

for any time $t > 0$. At the boundary Γ, the temperature is prescribed by

$$\theta(x, y, t)_\Gamma = \phi(x, y, t) \tag{3.70}$$

and the initial temperature distribution is

$$\theta(x, y, 0) = \psi(x, y) \tag{3.71}$$

Differential equation (3.69) may be expressed in a more compact form by introducing the differential operators

$$L_x \theta = \frac{\partial}{\partial x}\left[\kappa_1(x, y, t) \frac{\partial \theta}{\partial x}\right] \quad L_y \theta = \frac{\partial}{\partial y}\left[\kappa_2(x, y, t) \frac{\partial \theta}{\partial y}\right] \tag{3.72}$$

Substitution of (3.72) into Eq. (3.69) gives

$$\frac{\partial \theta}{\partial t} = L_x \theta + L_y \theta + f \tag{3.73}$$

To write a difference scheme, a difference grid is used, with the grid points having coordinates (x_i, y_j, t^n), where

$$x_i = i \Delta x \quad i = 0, 1, 2, \ldots, J$$
$$y_j = j \Delta y \quad j = 0, 1, 2, \ldots$$
$$t^n = n\tau \quad n = 0, 1, 2, \ldots$$

HEAT-CONDUCTION PROBLEMS

The function value at the grid point (x_i, y_j, t^n) is denoted by the indices; for example, $\theta(x_i, y_j, t^n) = \theta_{i,j}^n$. The differential operators L_x and L_y are approximated by difference operators δ_x and δ_y, respectively:

$$\delta_x \theta = \frac{1}{\Delta x} \left[a_{i+1,j} \frac{\theta_{i+1,j} - \theta_{i,j}}{\Delta x} - a_{i,j} \frac{\theta_{i,j} - \theta_{i-1,j}}{\Delta x} \right]$$
$$\delta_y \theta = \frac{1}{\Delta y} \left[b_{i,j+1} \frac{\theta_{i,j+1} - \theta_{i,j}}{\Delta y} - b_{i,j} \frac{\theta_{i,j} - \theta_{i,j-1}}{\Delta y} \right] \quad (3.74)$$

The coefficients $a_{i,j}$ and $b_{i,j}$ are chosen so that at any t ($t^n \leq t \leq t^{n+1}$) the operators L_x and L_y are of the second order of approximation, that is, $L_x \theta - \delta_x \theta = O((\Delta x)^2)$ and $L_y \theta - \delta_y \theta = O((\Delta y)^2)$. The following formulas may be recommended [6]:

$$a_{i,j} = \kappa_1(x_{i-1/2}, y_j, \bar{t}) \quad b_{i,j} = \kappa_2(x_i, y_{j-1/2}, \bar{t}) \quad (3.75a)$$

$$a_{i,j} = 0.5[\kappa_1(x_i, y_j, \bar{t}) + \kappa_1(x_{i-1}, y_j, \bar{t})]$$
$$b_{i,j} = 0.5[\kappa_2(x_i, y_j, \bar{t}) + \kappa_2(x_i, y_{j-1}, \bar{t})] \quad (3.75b)$$
$$\bar{t} = t^{n+1/2} = t^n + 0.5\tau$$

If κ_1 and κ_2 are constant, then

$$\delta_x \theta = \frac{\kappa_1}{(\Delta x)^2}(\theta_{i+1,j} - 2\theta_{i,j} + \theta_{i-1,j})$$

$$\delta_y \theta = \frac{\kappa_2}{(\Delta y)^2}(\theta_{i,j+1} - 2\theta_{i,j} + \theta_{i,j-1})$$

The following finite-difference schemes may be recommended for problem (3.69)–(3.71):

1. A longitudinal-transverse scheme. Along with the basic values of the desired grid function $\theta_{i,j}^n$ and $\theta_{i,j}^{n+1}$, an intermediate value $\theta_{i,j}^{n+1/2}$ is introduced. The transition from level t^n to level t^{n+1} is made in two stages, using steps of 0.5τ:

$$\theta_{i,j}^{n+1/2} = \theta_{i,j}^n + 0.5(\delta_x \theta^{n+1/2} + \delta_y \theta^n + f_{i,j}^{n+1/2})$$
$$\theta_{i,j}^{n+1} = \theta_{i,j}^{n+1/2} + 0.5(\delta_x \theta^{n+1/2} + \delta_y \theta^{n+1} + f_{i,j}^{n+1/2}) \quad (3.76)$$

The first of these equations is implicit in direction x and explicit in y; the second is explicit in x and implicit in y. Equation (3.76) must be supplemented with the boundary and initial conditions, which may be written in a difference form as

$$\theta_{i,j}^{n+1/2} = \bar{\phi}_{i,j} \quad \text{for } i=0, i=J$$
$$\theta_{i,j}^{n+1} = \phi_{i,j}^{n+1} \quad \text{for } j=0, j=\mathcal{J} \tag{3.77}$$

where $\bar{\phi}_{i,j} = 0.5(\phi_{i,j}^{n+1} - \phi_{i,j}^{n}) - 0.25\tau\delta_y(\phi^{n+1} - \phi^n)$

The meaning of the second boundary condition is quite clear. The first condition is chosen so that difference scheme (3.76)–(3.77) is of the second order of approximation.

For convenience in computer programming, difference scheme (3.76)–(3.77) is rewritten as

$$-F_{i,j}^n = \frac{a_{i,j}}{(\Delta x)^2}\theta_{i-1,j}^{n+1/2} - \left[\frac{a_{i+1,j}+a_{i,j}}{(\Delta x)^2} + \frac{2}{\tau}\right]\theta_{i,j}^{n+1/2} + \frac{a_{i+1,j}}{(\Delta x)^2}\theta_{i+1,j}^{n+1/2}$$

$$F_{i,j}^n = \frac{b_{i,j}}{(\Delta y)^2}\theta_{i,j-1}^n - \left[\frac{b_{i,j+1}+b_{i,j}}{(\Delta y)^2} - \frac{2}{\tau}\right]\theta_{i,j}^n + \frac{b_{i,j+1}}{(\Delta y)^2}\theta_{i,j+1}^n + f_{i,j}^{n+1/2}$$

$$\theta_{i,j}^{n+1/2} = \bar{\phi}_{0,j} \quad \theta_{J,j}^{n+1/2} = \bar{\phi}_{J,j} \quad i=1,2,3,\ldots,J-1 \tag{3.78}$$

$$-F_{i,j}^{n+1/2} = \frac{b_{i,j}}{(\Delta y)^2}\theta_{i,j-1}^{n+1} - \left[\frac{b_{i,j+1}+b_{i,j}}{(\Delta y)^2} + \frac{2}{\tau}\right]\theta_{i,j}^{n+1} + \frac{b_{i,j+1}}{(\Delta y)^2}\theta_{i,j+1}^{n+1}$$

$$F_{i,j}^{n+1/2} = \frac{a_{i,j}}{(\Delta x)^2}\theta_{i-1,j}^{n+1/2} - \left[\frac{a_{i+1,j}+a_{i,j}}{(\Delta x)^2} - \frac{2}{\tau}\right]\theta_{i,j}^{n+1/2} + \frac{a_{i+1,j}}{(\Delta x)^2}\theta_{i+1,j}^{n+1/2}$$
$$+ f_{i,j}^{n+1/2}$$

$$\theta_{i,0}^{n+1} = \phi_{i,0}^{n+1} \quad \theta_{i,\mathcal{J}}^{n+1} = \phi_{i,\mathcal{J}}^{n+1} \quad j=1,2,3,\ldots,\mathcal{J}-1 \tag{3.79}$$

Let $\theta_{i,j}^n$ be known. Then $F_{i,j}^n$ may be computed, and with $j=1$ fixed, Eq. (3.77) may be solved and $\theta_{i,1}^{n+1/2}$ determined using the factorization method. Next, fixing $j=2$, we may find $\theta_{i,2}^{n+1/2}$, and so on, until the $\theta_{i,j}^{n+1/2}$ are calculated at all the grid points.

After each row of (3.78) is found, each column of (3.79) may be successively solved in a similar way. For this purpose, with $i=1$ fixed, $\theta_{1,j}^{n+1}$ is found; then with $i=2$ fixed, $\theta_{2,j}^{n+1}$ is found; and so on until all the values of the desired function $\theta_{i,j}^{n+1}$ for the $(n+1)$th time step are obtained. The computation procedure is repeated, moving from the $(n+1)$th step to the $(n+2)$th step; i.e., the directions are alternated.

Stability and convergence tests of such difference schemes are usually performed by eliminating the intermediate values $\theta_{i,j}^{n+1/2}$. It may easily be seen that difference scheme (3.76)-(3.77) approximates the original problem with the second order of approximation $O(\tau^2 + (\Delta x)^2 + (\Delta y)^2)$. Stability analysis [6] reveals that if the coefficients κ_1 and κ_2 are independent of time, that is, $\kappa_1 = \kappa_1(x, y)$ and $\kappa_2 = \kappa_2(x, y)$, then scheme (3.76)-(3.77) is absolutely stable. When κ_1 and κ_2 are functions of time, the absolute stability of the difference scheme cannot be proved; however, numerical experimentation with some example problems reveals that for $\tau \approx h$ [$h = \min(\Delta x, \Delta y)$], the computations remain stable.

2. Stabilization procedure. The stabilization procedure may be described [12] as follows:

$$\theta_{i,j}^{n+1/2} - 0.5\tau\delta_x\theta^{n+1/2} = \delta_x\theta^n + \delta_y\theta^n + f_{i,j}^{n+1/2}$$
$$\bar{\theta}_{i,j}^{n+1} - 0.5\tau\delta_y\bar{\theta}^{n+1} = \theta_{i,j}^{n+1/2} \qquad (3.80)$$
$$\theta_{i,j}^{n+1} = \theta_{i,j}^n + \tau\bar{\theta}_{i,j}^{n+1}$$

where $\theta_{i,j}^{n+1/2}$ and $\bar{\theta}_{i,j}^{n+1}$ are auxiliary quantities that permit problem (3.69)-(3.71) to be reduced to a succession of simple problems. The first and second difference equations are implicit, and they may be solved as before by the factorization procedure; the third equation is explicit.

A stability analysis [6, 12] for the scheme reveals that when the coefficients κ_1 and κ_2 are independent of time and when the solution θ and right-hand side f are sufficiently smooth, then difference scheme (3.80) is absolutely stable and gives the solution of problem (3.69)-(3.71) with an error of order $O(\tau^2 + (\Delta x)^2 + (\Delta y)^2)$.

3. The predictor-corrector scheme. The main idea behind the method is that in proceeding from the nth time level to the $(n+1)$th level, the solution of problem (3.69)-(3.71) is divided into two stages. First, from a scheme with first-order accuracy, an approximate solution is determined at time $t^{n+1/2} = t^n + 0.5\tau$. Then, for the whole interval $t^n \leq t \leq t^{n+1}$, the initial equation is solved with the second order of approximation. The second part is used as a corrector. A very essential factor is the use, in constructing the correction, of a "rough" solution found at $t = t^{n+1/2}$ with the aid of the predictor.

The predictor-corrector scheme may be written [22] as

$$\theta_{i,j}^{n+1/4} = \theta_{i,j}^n + 0.5\tau\delta_x\theta^{n+1/4}$$
$$\theta_{i,j}^{n+1/2} = \theta_{i,j}^{n+1/4} + 0.5\tau\delta_y\theta^{n+1/2} \qquad (3.81)$$
$$\theta_{i,j}^{n+1} = \theta_{i,j}^n + \tau(\delta_x\theta^{n+1/2} + \delta_y\theta^{n+1/2} + f_{i,j}^{n+1/2})$$

The first two equations represent the predictor. They both are implicit, and the solution of either may be obtained by the factorization method. The last equation is the corrector; it is explicit.

Difference scheme (3.81) may be demonstrated to be absolutely stable provided all the coefficients of the initial equation (3.69) are independent of time; the scheme has an error of order $O(\tau^2 + (\Delta x)^2 + (\Delta y)^2)$.

It may easily be seen that all the difference schemes considered for problem (3.69)–(3.71) with $L_x L_y = L_y L_x$ (i.e., with permutable operators), with the right-hand function $f = 0$, and based on the fractional-step method, are equivalent. They thus serve only as different computation schemes. The longitudinal-transverse scheme seems most advantageous, since it requires the smallest amount of computer storage. For the calculation of $\theta_{i,j}^{n+1}$ in this scheme, it is not necessary to store values of $\theta_{i,j}^n$, as in the other schemes.

For nonhomogeneous problems $(f \neq 0)$, all the above difference schemes that are of the second order of approximation in τ, Δx, and Δy lead to results of order $O(\tau^2 + (\Delta x)^2 + (\Delta y)^2)$. It is difficult to recommend areas of most efficient application of these difference schemes, since the subject has not been studied sufficiently. However, the very fact that three different procedures may be used to solve the same problem allows such problems to be approached with much greater confidence.

The convergence of all three difference methods is ensured when the operators L_x and L_y are independent of time. Otherwise, convergence cannot as yet be proved.

There is a two-cycle modification of the longitudinal-transverse method for which stability can be proved when L_x and L_y are time dependent. Consider the system of difference equations

$$\theta_{i,j}^{n-1/2} - 0.5\tau\delta_x \theta^{n-1/2} = \theta_{i,j}^{n-1} + 0.5\tau\delta_x \theta^{n-1}$$

$$\bar{\theta}_{i,j}^n - 0.5\tau\delta_y \bar{\theta}^n = \theta_{i,j}^{n-1/2} + 0.5\tau\delta_y \theta^{n-1/2}$$

$$\theta_{i,j}^n = \bar{\theta}_{i,j}^n + 2\tau f_{i,j}^n \qquad (3.82)$$

$$\theta_{i,j}^{n+1/2} - 0.5\tau\delta_y \theta^{n+1/2} = \theta_{i,j}^n + 0.5\tau\delta_y \theta^n$$

$$\theta_{i,j}^{n+1} - 0.5\tau\delta_x \theta^{n+1} = \theta_{i,j}^{n+1/2} + 0.5\tau\delta_x \theta^{n+1/2}$$

Difference scheme (3.82) allows us to go from the $(n-1)$th level immediately to the $(n+1)$th level. The calculation is divided into two cycles. In the first cycle, an ordinary longitudinal-transverse scheme is used,

HEAT-CONDUCTION PROBLEMS   91

and in the second cycle the same scheme is used but in the reverse order. During the calculation, the operators L_x and L_y are approximated over the interval $t^{n-1} \leqslant t \leqslant t^{n+1}$; that is, in formulas (3.75), $\bar{t} = t^n$ is used in place of $\bar{t} = t^{n+1/2}$ throughout.

It may be demonstrated [12] that difference scheme (3.82) approximates differential equation (3.69) with approximation order $O(\tau + (\Delta x)^2 + (\Delta y)^2)$ and that it is absolutely stable, if $L_x(t) \geqslant 0$ and $L_y(t) \geqslant 0$.

3.8 DIFFERENCE METHODS APPLICABLE TO THREE-DIMENSIONAL HEAT-CONDUCTION PROBLEMS

The difficulties involved in shifting from one-dimensional proclems to two-dimensional problems are compounded when one attempts to solve three-dimensional problems. The number of difference equations and, accordingly, the number of unknown quantities are then proportional to h^{-3}. This results in an increase in the computation time of more than an order, as compared with the two-dimensional case.

Consider a three-dimensional heat-conduction problem written in operator form as

$$\frac{\partial \theta}{\partial t} = L_x \theta + L_y \theta + L_z \theta + f \tag{3.83}$$

where

$$L_x \theta = \frac{\partial}{\partial x} \kappa_1(x, y, z, t) \frac{\partial \theta}{\partial x}$$

$$L_y \theta = \frac{\partial}{\partial y} \kappa_2(x, y, z, t) \frac{\partial \theta}{\partial y}$$

$$L_z \theta = \frac{\partial}{\partial z} \kappa_3(x, y, z, t) \frac{\partial \theta}{\partial z}$$

are the differential operators. The solution $\theta(x, y, z, t)$ is sought inside the parallelepiped $0 \leqslant x \leqslant l_x$, $0 \leqslant y \leqslant l_y$, $0 \leqslant z \leqslant l_z$, with boundary conditions

$$\theta_\Gamma = \phi(x, y, z, t) \tag{3.84}$$

In addition, the temperature distribution in the parallelepiped at the initial time $t = 0$ is known:

$$\theta(x, y, z, 0) = \psi(x, y, z) \tag{3.85}$$

$\theta(x, y, z, t)$ is to be determined for $t > 0$.

For the numerical solution of problem (3.83)–(3.85), a difference grid using steps $\Delta x = l_x/J$, $\Delta y = l_y/\mathcal{I}$, $\Delta z = l_z/K$, and τ is taken for variables x, y, z, and t, respectively. The operators L_x, L_y, and L_z are approximated to the second order of accuracy as in the two-dimensional case:

$$\delta_x \theta = \frac{1}{(\Delta x)^2} [a_{i+1,j,k}(\theta_{i+1,j,k} - \theta_{i,j,k}) - a_{i,j,k}(\theta_{i,j,k} - \theta_{i-1,j,k})]$$

$$\delta_y \theta = \frac{1}{(\Delta y)^2} [b_{i,j+1,k}(\theta_{i,j+1,k} - \theta_{i,j,k}) - b_{i,j,k}(\theta_{i,j,k} - \theta_{i,j-1,k})] \tag{3.86}$$

$$\delta_z \theta = \frac{1}{(\Delta z)^2} [c_{i,j,k+1}(\theta_{i,j,k+1} - \theta_{i,j,k}) - c_{i,j,k}(\theta_{i,j,k} - \theta_{i,j,k-1})]$$

where a, b, and c may be determined from the formulas

$$a_{i,j,k} = \kappa_1(x_{i-1/2}, y_j, z_k, \bar{t}) \quad b_{i,j,k} = \kappa_2(x_i, y_{j-1/2}, z_k, \bar{t})$$
$$c_{i,j,k} = \kappa_3(x_i, y_j, z_{k-1/2}, \bar{t}) \quad \bar{t} = t^n + 0.5\tau \tag{3.87}$$

The classical explicit scheme

$$\frac{\theta_{i,j,k}^{n+1} - \theta_{i,j,k}^n}{\tau} = \delta_x \theta^n + \delta_y \theta^n + \delta_z \theta^n + f_{i,j,k} \tag{3.88}$$

which approximates Eq. (3.83) with an error of order $O(\tau + (\Delta x)^2 + (\Delta y)^2 + (\Delta z)^2)$, is known to be stable only for $\tau \leq h^2/6$ where $h = \min(\Delta x, \Delta y, \Delta z)$. The classical implicit difference scheme

$$\frac{\theta_{i,j,k}^{n+1} - \theta_{i,j,k}^n}{\tau} = \delta_x \theta^{n+1} + \delta_y \theta^{n+1} + \delta_z \theta^{n+1} + f_{i,j,k} \tag{3.89}$$

which approximates Eq. (3.83) with the same order $O(\tau + (\Delta x)^2 + (\Delta y)^2 + (\Delta z)^2)$, is absolutely stable, but it requires solution at each level of

HEAT-CONDUCTION PROBLEMS

an equation system involving a seven-diagonal matrix. The disadvantages of Eqs. (3.88) and (3.89) are quite clear.

Scheme (3.89) is sometimes used when a steady (independent of time) solution of problem (3.83)–(3.85) is sought. The parameter τ is then considered as iterational, and the solution of the difference equations is found with one of the procedures discussed in Sec. 2.7.

For the solution of unsteady problems (as well as for steady ones), difference schemes based on the fractional-step method seem most useful. It should, however, be noted that it may not always be possible to extend two-dimensional schemes, of the type considered above, to three-dimensional problems. Consider the following finite-difference procedures.

1. One scheme that may be used for three-dimensional problems is the stabilization procedure. The numerical algorithm for this procedure may be written as

$$\begin{aligned}
\theta_{i,j,k}^{n+1/3} - 0.5\tau\delta_x\theta^{n+1/3} &= \delta_x\theta^n + \delta_y\theta^n + \delta_z\theta^n + f_{i,j,k}^{n+1/2} \\
\theta_{i,j,k}^{n+2/3} - 0.5\tau\delta_y\theta^{n+2/3} &= \theta_{i,j,k}^{n+1/3} \\
\bar{\theta}_{i,j,k}^{n+1} - 0.5\tau\delta_z\bar{\theta}^{n+1} &= \theta_{i,j,k}^{n+2/3} \\
\theta_{i,j,k}^{n+1} &= \theta_{i,j,k}^n + \tau\bar{\theta}_{i,j,k}^{n+1}
\end{aligned} \quad (3.90)$$

It may easily be verified that the stabilization procedure in the case of a sufficiently smooth solution is of the second order of accuracy and is absolutely stable if the coefficients of the initial differential equation are independent of time and the operators L_x, L_y, and L_z are commutative [12]. $\theta_{i,j,k}^{n+1}$ may be evaluated with $\theta_{i,j,k}^n$ known from the successive solution of three implicit difference equations and one explicit equation. The implicit equations are usually solved by the factorization method.

2. The predictor-corrector method may also be extended to the three-dimensional case. Here the predictor does not consist of two implicit equations, as in the two-dimensional case, but of the equations

$$\begin{aligned}
\theta_{i,j,k}^{n+1/4} - \theta_{i,j,k}^n &= 0.5\tau\delta_x\theta^{n+1/4} \\
\theta_{i,j,k}^{n+1/2} - \theta_{i,j,k}^{n+1/4} &= 0.5\tau\delta_y\theta^{n+1/2} \\
\bar{\theta}_{i,j,k}^{n+1} - \theta_{i,j,k}^{n+1/2} &= 0.5\tau\delta_z\bar{\theta}^{n+1} \\
\theta_{i,j,k}^{n+1} &= \theta_{i,j,k}^n + \tau(\delta_x\bar{\theta}^{n+1} + \delta_y\bar{\theta}^{n+1} + \delta_z\bar{\theta}^{n+1} + f_{i,j,k}^{n+1/2})
\end{aligned} \quad (3.91)$$

For sufficiently smooth solutions, this method also gives a second-order approximation for all discrete parameters. Unfortunately, this difference procedure may be stable only when the commutative operators L_x, L_y, and L_z are constant with respect to time. The question of the stability of the two schemes has not been answered as yet, although numerical experiments have revealed that for some problems the schemes are stable for $\tau \approx h$.

3. The two-cycle difference scheme based on the splitting procedure is of interest in solving these problems [12]. This scheme, which was discussed previously, may be extended to the three-dimensional case. The numerical algorithm is of the form

$$\theta_{i,j,k}^{n-2/3} - 0.5\tau\delta_x\theta^{n-2/3} = \theta_{i,j,k}^{n-1} + 0.5\tau\delta_x\theta^{n-1}$$

$$\theta_{i,j,k}^{n-1/3} - 0.5\tau\delta_y\theta^{n-1/3} = \theta_{i,j,k}^{n-2/3} + 0.5\tau\delta_y\theta^{n-2/3}$$

$$\bar{\theta}_{i,j,k}^{n} - 0.5\tau\delta_z\bar{\theta}^{n} = \theta_{i,j,k}^{n-1/3} + 0.5\tau\delta_z\theta^{n-1/3}$$

$$\theta_{i,j,k}^{n} = \bar{\theta}_{i,j,k}^{n} + 2\tau f_{i,j,k}^{n} \qquad (3.92)$$

$$\theta_{i,j,k}^{n+1/3} - 0.5\tau\delta_z\theta^{n+1/3} = \theta_{i,j,k}^{n} + 0.5\tau\delta_z\theta^{n}$$

$$\theta_{i,j,k}^{n+2/3} - 0.5\tau\delta_y\theta^{n+2/3} = \theta_{i,j,k}^{n+1/3} + 0.5\tau\delta_y\theta^{n+1/3}$$

$$\theta_{i,j,k}^{n+1} - 0.5\tau\delta_x\theta^{n+1} = \theta_{i,j,k}^{n+2/3} + 0.5\tau\delta_x\theta^{n+2/3}$$

First, a system of difference equations is solved in the interval $t^{n-1} \leqslant t \leqslant t^n$ for i, j, k; then, the same system is used in the reverse order in the interval $t^n \leqslant t \leqslant t^{n+1}$ for k, j, i.

This difference scheme is of the second order of approximation in τ, Δx, Δy, and Δz and is absolutely stable even when the operators L_x, L_y, and L_z are noncommutative and depend on time.

Although the formulas are somewhat complex, the numerical algorithm for the scheme is very simple and easy to program on a high-speed computer. The advantages of the scheme make it highly promising for the solution of very general multidimensional heat-conduction problems.

3.9 NONLINEAR HEAT-CONDUCTION PROBLEMS

In previous sections, we discussed linear heat-conduction problems with the coefficients and right-hand sides independent of temperature. But real

HEAT-CONDUCTION PROBLEMS

situations always involve temperature-dependent coefficients that should be taken into account. The temperature dependence of heat capacity and thermal conductivity has an especially great effect on heat transfer in high-temperature processes. Such temperature-dependent coefficients lead to nonlinear heat-conduction problems. Heat transfer in bodies with nonlinear sources and nonlinear boundary conditions (e.g., radiant heat transfer) are governed by nonlinear equations.

The finite-difference techniques are practically the only methods applicable to nonlinear problems. However, no general and reliable methods are as yet available for analysis of the convergence of difference schemes for nonlinear equations. Such analysis is generally possible only for linearized systems; the results may then be used for nonlinear equations. This approach is often used to determine the convergence conditions for nonlinear equations.

In practical computations, an iterative procedure is carried out at every time step to determine more accurate coefficients, which are solution dependent. In such cases, the stability of the computation may easily be controlled through the number of iterations required for the desired accuracy. For example, if the number of iterations for a given time level exceeds the admissible number N_{max}, the time step τ is decreased by a factor of two, and the computation is repeated for the same time level but with decreased time step $\tau' = \tau/2$. If the required accuracy is achieved with fewer iterations than a prescribed number N_{min} (usually, $N_{min} = 2$ or 3), the new time level is computed with $\tau^{n+1} = 1.3\tau^n$. When the number of iterations Σ at a given time level is within $N_{min} \leq \Sigma \leq N_{max}$, the new $(n+1)$th time level is computed with $\tau^{n+1} = 0.8\tau^n$.

Consider the very simple heat-conduction problem

$$\frac{\partial \theta}{\partial t} = \frac{\partial}{\partial x}\left[\kappa(\theta)\frac{\partial \theta}{\partial x}\right] + f(\theta) \qquad 0 \leq x \leq 1 \qquad t > 0$$

$$\theta(x, 0) = \psi(x) \qquad \theta(0, t) = \psi_1(t) \qquad \theta(1, t) = \psi_2(t)$$

(3.93)

where $\kappa(\theta) > 0$

It should be pointed out immediately that explicit schemes are not desirable for such problems since their stability condition $\tau \leq h^2 / 2 \max_\theta \kappa(\theta)$ requires a very small time step. More advantageous in this case is the use of absolutely stable implicit schemes. Consider the following two types of implicit difference schemes for problem (3.93), both commonly used in practice:

$$\frac{\theta_i^{n+1} - \theta_i^n}{\tau} = \frac{1}{h}\left[a_{i+1}(\theta^n)\frac{\theta_{i+1}^{n+1} - \theta_i^{n+1}}{h} - a_i(\theta^n)\frac{\theta_i^{n+1} - \theta_{i-1}^{n+1}}{h}\right] + f(\theta^n)$$
(3.94)

and

$$\frac{\theta_i^{n+1} - \theta_i^n}{\tau} = \frac{1}{h}\left[a_{i+1}(\theta^{n+1})\frac{\theta_{i+1}^{n+1} - \theta_i^{n+1}}{h} - a_i(\theta^{n+1})\frac{\theta_i^{n+1} - \theta_{i-1}^{n+1}}{h}\right]$$
$$+ f(\theta^{n+1}) \quad (3.95)$$

where $a_i(\theta) = \kappa[0.5(\theta_i + \theta_{i+1})]$

Let us compare the two schemes. Each scheme has an error of $O(\tau + h^2)$. Both are absolutely stable. Scheme (3.94) is linear for functions $\theta_{i,j}^{n+1}$ at time level t^{n+1}. Functions $\theta_{i,j}^{n+1}$ at time level t^n may be evaluated from $\theta_{i,j}^n$ by the factorization method. Since scheme (3.94) is absolutely stable, the time step τ is chosen strictly on the basis of accuracy considerations. Scheme (3.95) is nonlinear for functions $\theta_{i,j}^{n+1}$, and its solution requires iteration. The iteration procedure may be arranged as follows:

$$\frac{\theta_i^{s+1} - \theta_i^n}{\tau} = \frac{1}{h}\left[a_{i+1}(\theta^s)\frac{\theta_{i+1}^{s+1} - \theta_i^{s+1}}{h} - a_i(\theta^s)\frac{\theta_i^{s+1} - \theta_{i-1}^{s+1}}{h}\right] + f(\theta^s)$$
(3.96)

Iteration difference scheme (3.96) is linear for θ^{s+1}. Therefore, θ^{s+1} may be found at every iteration by the factorization method. As an initial approximation for the function θ, the value of the function at the preceding time step, that is, $\theta^{s=0} = \theta^n$ is ordinarily used. Iteration procedures for most practical coefficients κ and f are convergent. Two or three iterations appear enough under normal conditions. When scheme (3.96) is used for computation, either the number of iterations or the convergence accuracy must be prescribed; i.e., the following condition must be satisfied:

$$\frac{\max_i |\theta_i^{s+1} - \theta_i^s|}{\max_i |\theta_i^{s+1}|} \leq \epsilon$$
(3.97)

If such is the case, then $\theta_{i,j}^{s+1} = \theta_{i,j}^{n+1}$ is assumed.

HEAT-CONDUCTION PROBLEMS

The disadvantage of scheme (3.95) is that it requires twice as much storage as scheme (3.94), since the computation of $\theta_{i,j}^{n+1}$ requires the storage of both $\theta_{i,j}^{n}$ and $\theta_{i,j}^{s}$. Moreover, the use of (3.95) for $\theta_{i,j}^{n+1}$ requires a certain number of iterations for each time level. Since both schemes are absolutely stable and of the same order of approximation, one would expect that scheme (3.94) is better than (3.95). This is not the case, however. Numerical experiments and practice have revealed that the actual accuracy of difference scheme (3.95) is considerably higher. Scheme (3.95), employing a larger time step, requires a smaller total number of operations to produce the same accuracy, despite the necessity for iterations.

There are schemes with second-order accuracy, with respect to space and time, for the solution of nonlinear heat-conduction problems. For example, problem (3.93) may be solved by

$$\frac{\theta_i^{n+1} - \theta_i^n}{\tau} = \frac{1}{2}(\delta_x \theta^{n+1} + \delta_x \theta^n) + f\left(\frac{\theta_i^{n+1} + \theta_i^n}{2}\right) \tag{3.98}$$

where

$$\delta_x \theta = \frac{1}{h}\left[a_{i+1}(\theta)\frac{\theta_{i+1} - \theta_i}{h} - a_i(\theta)\frac{\theta_i - \theta_{i-1}}{h}\right]$$

This scheme is of approximation order $O(\tau^2 + h^2)$. Such schemes are not, however, of the monotone type, and at large temperature gradients they lead to false solutions. They may be improved by using a sufficiently small time step.

We have considered a one-dimensional case. All the above difference schemes may be extended to two and more dimensions. Here, implicit difference schemes based on the fractional-step method may be recommended. Among these, the two-cycle procedure based on the splitting method seems most promising, since it is absolutely stable even when the operators L_x, L_y, and L_z are noncommutative and time dependent.

When such schemes are used to difference the differential operators L_x, L_y, and L_z, their dependence on θ should be taken into account. The following version is recommended:

$$L_x \theta \approx \delta_x \theta = \frac{1}{\Delta x}\left(a_{i+1,j,k}\frac{\theta_{i+1,j,k} - \theta_{i,j,k}}{\Delta x} - a_{i,j,k}\frac{\theta_{i,j,k} - \theta_{i-1,j,k}}{\Delta x}\right) \tag{3.99}$$

where

$$a_{i,j,k} = \kappa_1 \left[x_{i-1/2}, y_j, z_k, t^{n+1/2}, \frac{\theta_{i+1,j,k} + \theta_{i,j,k}}{2} \right]$$

The operators L_y and L_z are approximated in a similar manner. In all cases, the solution requires iterations for every integral time step.

Chapter 4

Convective Heat Transfer

4.1 CONVECTION EQUATIONS; BOUNDARY CONDITIONS

So far, we have considered only problems of heat propagation in immovable media. Studies of heat transfer in fluids are somewhat more complicated, since the effects of fluid motion on heat transfer must be taken into account. (In what follows, the term *fluids* will be used to denote both gases and liquids, unless otherwise indicated.) In solids, heat is transferred only by molecular heat conduction and radiation; in fluids, convection contributes much to heat transfer. The convective motion may be induced either by external forces (forced convection) or by density differences resulting from local heating in the gravity field (free convection).

Studies of convective heat transfer are reduced to the solution of nonlinear systems of partial differential equations involving the conservation principles of energy, momentum, and mass, state equations, etc. The complete formulation of such a problem is rather difficult, and even an approximate solution cannot be obtained in many cases. Therefore, convective heat-transfer problems are usually studied after certain simplifying assumptions are made, to allow analysis by the available means.

The approximation that is most widely used involves the assumption of an incompressible fluid with constant properties independent of temperature (except density). Variations of density with temperature are accounted for only in the buoyancy expression. The density is assumed to be independent of pressure. Temperature variations resulting from heat generation and viscous dissipation are neglected. This approach to convective heat transfer was originally suggested by Overbeak and later developed in the works of Boussinesque. Numerous analytical and experimental works verify the applicability of the approach to many practical problems.

The system of equations for convective heat and mass transfer in the Boussinesque approximation may be formulated [13, 14] as follows:

$$\frac{\partial \bar{v}}{\partial t} + (\bar{v}\nabla)\bar{v} = -\frac{1}{\rho_0}\nabla P + \nu \nabla^2 \bar{v} - g\beta(T - T_0)$$

$$\frac{\partial T}{\partial t} + \bar{v}\nabla T = a\nabla^2 T \qquad (4.1)$$

$$\text{div } \bar{v} = 0 \qquad \rho - \rho_0 = -\rho_0 \beta(T - T_0)$$

where ρ_0 = mean (reference) density at constant temperature T_0
$\beta = -1/\rho_0 (\partial \rho/\partial T)_{T_0}$ = thermal expansion coefficient of fluid
ν = kinematic viscosity
a = thermal diffusivity

For a two-dimensional case in Cartesian coordinates, system (4.1) may be reduced to the following dimensionless form:

$$\frac{\partial u}{\partial t} + u\frac{\partial u}{\partial x} + v\frac{\partial u}{\partial y} + \frac{\partial P'}{\partial x} = \frac{1}{\text{Re}}\left(\frac{\partial^2 u}{\partial x^2} + \frac{\partial^2 u}{\partial y^2}\right) - \frac{\text{Gr}\theta}{\text{Re}^2}\sin\varphi$$

$$\frac{\partial v}{\partial t} + u\frac{\partial v}{\partial x} + v\frac{\partial v}{\partial y} + \frac{\partial P'}{\partial y} = \frac{1}{\text{Re}}\left(\frac{\partial^2 v}{\partial x^2} + \frac{\partial^2 v}{\partial y^2}\right) + \frac{\text{Gr}\theta}{\text{Re}^2}\cos\varphi$$

$$\frac{\partial u}{\partial x} + \frac{\partial v}{\partial y} = 0 \qquad (4.2)$$

$$\frac{\partial \theta}{\partial t} + u\frac{\partial \theta}{\partial x} + v\frac{\partial \theta}{\partial y} = \frac{1}{\text{Re Pr}}\left(\frac{\partial^2 \theta}{\partial x^2} + \frac{\partial^2 \theta}{\partial y^2}\right)$$

where u, v = velocity projections on x, y axes, respectively
P' = pressure deviation from static value
θ = temperature
t = time

As references for dimensionless quantities, we select the characteristic length L, velocity v_1, temperature difference $\Delta T = T_1 - T_0$, and time L/v_1.

System (4.2) contains three dimensionless quantities, namely, the Prandtl number $\mathrm{Pr} = \nu/a$, Grashof number $\mathrm{Gr} = g\beta L^3 \Delta T/\nu^2$, and Reynolds number $\mathrm{Re} = v_1 L/\nu$. Here g is the acceleration at the angle φ from the vertical axis resulting from an external force. In some cases, the Rayleigh number $\mathrm{Ra} = \mathrm{GrPr}$ may be usefully employed to describe the heat transfer.

When the external fluid flow has velocity v_1, system (4.2) describes a simultaneous free and forced convection problem. Pure forced convection occurs when $\mathrm{Gr} = 0$. Free convection is governed by the same system with one modification: for the reference velocity v_1 we assume ν/L, which implies $\mathrm{Re} = 1$. L^2/ν is used as the time reference.

For the case of flat-plate flow, the elimination of pressure from (4.2) and the introduction of the stream function ψ

$$\frac{\partial \psi}{\partial x} = -v \qquad \frac{\partial \psi}{\partial y} = u \qquad (4.3)$$

may be useful. The continuity equation $\mathrm{div}\, \bar{v} = 0$ is then satisfied automatically. For convenience in constructing the numerical algorithm, the vorticity function ω may also be used. These transformations reduce (4.2) to the form

$$\frac{\partial \theta}{\partial t} + u\frac{\partial \theta}{\partial x} + v\frac{\partial \theta}{\partial y} = \frac{1}{\mathrm{Re}\,\mathrm{Pr}}\left(\frac{\partial^2 \theta}{\partial x^2} + \frac{\partial^2 \theta}{\partial y^2}\right)$$

$$\frac{\partial \omega}{\partial t} + u\frac{\partial \omega}{\partial x} + v\frac{\partial \omega}{\partial y} = \frac{1}{\mathrm{Re}}\left(\frac{\partial^2 \omega}{\partial x^2} + \frac{\partial^2 \omega}{\partial y^2}\right) + \frac{\mathrm{Gr}}{\mathrm{Re}^2}\left(\frac{\partial \theta}{\partial x}\sin\varphi + \frac{\partial \theta}{\partial y}\cos\varphi\right) \quad (4.4)$$

$$\frac{\partial^2 \psi}{\partial x^2} + \frac{\partial^2 \psi}{\partial y^2} = -\omega \qquad u = \frac{\partial \psi}{\partial y} \qquad v = -\frac{\partial \psi}{\partial x}$$

Since $\mathrm{div}\, \bar{v} = 0$, the system (4.4) may be written in a divergence form as

$$\frac{\partial \theta}{\partial t} = \frac{\partial}{\partial x}\left(\frac{1}{\mathrm{Re}\,\mathrm{Pr}}\frac{\partial \theta}{\partial x} - u\theta\right) + \frac{\partial}{\partial y}\left(\frac{1}{\mathrm{Re}\,\mathrm{Pr}}\frac{\partial \theta}{\partial y} - v\theta\right) \qquad (4.5)$$

$$\frac{\partial \omega}{\partial t} = \frac{\partial}{\partial x}\left(\frac{1}{\text{Re}}\frac{\partial \omega}{\partial x} - u\omega\right) + \frac{\partial}{\partial y}\left(\frac{1}{\text{Re}}\frac{\partial \omega}{\partial y} - v\omega\right)$$

$$+ \frac{\text{Gr}}{\text{Re}^2}\left(\frac{\partial \theta}{\partial x}\sin\varphi + \frac{\partial \theta}{\partial y}\cos\varphi\right) \quad (4.5)$$
cont.

$$\frac{\partial^2 \psi}{\partial x^2} + \frac{\partial^2 \psi}{\partial y^2} = -\omega$$

The numerical methods are based upon the solution of the above (θ, ω, ψ) Boussinesque system. The solution requires the formulation of the boundary conditions for the vorticity function ω. Indeed, with such a formulation of the problem, the vorticity function ω is defined only inside the region where the solution is sought. The solution of this (θ, ω, ψ) system is therefore only conformable for the given boundary conditions of velocity (or, equivalently, for stream function ψ) and temperature. However, the formal numerical solution does require that the boundary conditions for the function ω be given.

This difficulty may be overcome in different ways. One of the most common is as follows: The values of ω are found approximately at the boundary by using the computation for that ψ determined at the grid points in the vicinity of the boundary. Formulas for the approximate calculation of the boundary values of vorticity ω may be obtained either by Taylor-series expansion of the stream function ψ near the boundary points, or from the Poisson equation for the stream function with the assumption that it is valid at the boundary as well. In both cases, the boundary conditions must be taken into account.

Consider the simple case in which the boundary of the region coincides with the x axis, i.e., with $y = 0$ (in Cartesian coordinates), and the region of computation is above that line. A regular grid is used, with grid points having the coordinates (x_i, y_j, t^n), $x_i = i\,\Delta x$, $y_j = j\,\Delta y$, $t^n = n\tau$. The boundary conditions for the stream functions are then

$$\frac{\partial \psi}{\partial x} = f_1(x, t) \qquad \frac{\partial \psi}{\partial y} = f_2(x, t) \quad (4.6)$$

That is, the velocity components are given. A very simple computational formula for approximating $\omega_{i,0}$ is then

$$\omega_{i,0} = \frac{2(\psi_{i,0} - \psi_{i,1})}{(\Delta y)^2} + \frac{2f_2(x_i, t^n)}{\Delta y} - \left(\frac{\partial f_1}{\partial x}\right)_i^n + O(\Delta y) \quad (4.7)$$

CONVECTIVE HEAT TRANSFER

With $f_1 = f_2 = 0$, the condition of Tom is obtained. The major disadvantage of (4.7) is a low order of approximation in the region of the highest gradient of vorticity. If a strong instability is caused by this condition, a decrease in the time step is required.

A more accurate boundary condition for ω may be obtained from the two forms that follow. The first relates the vorticity at the boundary with the values at the points adjacent to the boundary [15]:

$$\omega_{i,0} = \frac{3}{(\Delta y)^2}[\psi_{i,0} - \psi_{i,1} + \Delta y\, f_2(x_i, t^n)] - \frac{3}{2}\left(\frac{\partial f_1}{\partial x}\right)_i - \frac{\Delta y}{2}\left(\frac{\partial^2 f_1}{\partial x^2}\right)_i$$

$$+ \frac{1}{2}\omega_{i,1} + O((\Delta y)^2) \quad (4.8)$$

In the second, ω is expressed only in terms of the stream functions:

$$\omega_{i,0} = \frac{7\psi_{i,0} - 8\psi_{i,1} + \psi_{i,2}}{2(\Delta y)^2} + \frac{3}{\Delta y}f_2(x_i, t^n) - \left(\frac{\partial f_1}{\partial x}\right)_i + O((\Delta y)^2) \quad (4.9)$$

The first formula is known as the *condition of Woods*, and the second was suggested by T. V. Kuskova. Formula (4.9) is more useful in practice since, unlike (4.8), it relates the calculation of the boundary condition for vorticity with stream functions only, and these may be obtained more quickly than the vorticity functions.

One additional formula of the third order could be suggested for boundary vorticities, but we shall limit ourselves to formulas (4.7)–(4.9).

An alternative approach to the solution of the (θ, ω, ψ) system, in which the boundary values of ω are not used, may give better results. In the formulation of convective heat-transfer problems, the boundary values of ψ and $\partial\psi/\partial n$ (a normal derivative) are usually given. The relation $\psi = \partial\psi/\partial n = 0$ implies the no-slip condition. The essence of the method involves satisfying of the stream-function boundary condition at every time level. This is achieved by suitable corrections of ψ at the boundary points.

The computation procedure for $\theta_{i,j}^{n+1}$, $\omega_{i,j}^{n+1}$, and $\psi_{i,j}^{n+1}$ may be written as follows. Let us select, for the numerical solution of (4.4), a difference scheme composed of three difference equations, namely, those for stream function, vorticity, and temperature. The values of $\theta_{i,j}^n$, $\omega_{i,j}^n$, and $\psi_{i,j}^n$ at the nth time level are assumed known. Then:

1. From the stream-function difference equation, the $\omega_{i,j}^{n+1}$'s at the points adjacent to the boundary of the differencing grid (at the internal grid

points, one step from the boundary) are evaluated. For this purpose, the following formula may be used:

$$\omega_{i,j}^{n+1} = -\frac{\psi_{i+1,j}^n - 2\psi_{i,j}^n + \psi_{i-1,j}^n}{(\Delta x)^2} - \frac{\psi_{i,j+1}^n - 2\psi_{i,j}^n + \psi_{i,j-1}^n}{(\Delta y)^2} \quad (4.10)$$

2. These $\omega_{i,j}^{n+1}$'s are assumed to be the boundary values and, with the vorticity equations solved, the $\omega_{i,j}^{n+1}$'s in the remainder of the region are evaluated.
3. The values of stream function $\psi_{i,j}^{n+1}$ inside the main region are found from the stream-function difference equation with the assumption that the $\psi_{i,j}^n$'s are given at the boundary.
4. The values of the stream function at the boundary must be corrected for the known conditions of $\partial\psi/\partial n$. For example, suppose $\psi = 0$ and $\partial\psi/\partial n = 0$ are assumed at the boundary. If the three-point formula of second-order accuracy

$$\left(\frac{\partial \psi}{\partial n}\right)_0 = -\frac{3\psi_0 - 4\psi_1 + \psi_2}{2h} + O(h^2)$$

is used, satisfying the above conditions requires that $\psi_1^{n+1} = 0.25\psi_2^{n+1}$. A four-point approximation of third-order accuracy,

$$\left(\frac{\partial \psi}{\partial n}\right)_0 = -\frac{1}{6h}(11\psi_0 - 18\psi_1 + 9\psi_2 - 2\psi_3) + O(h^3)$$

may also be used. Then

$$\psi_1^{n+1} = \frac{1}{2}\psi_2^{n+1} - \frac{1}{9}\psi_3^{n+1} \quad \text{since} \quad \psi_0 = \left(\frac{\partial \psi}{\partial n}\right)_0 = 0$$

5. Finally, the temperature difference equation is used to determine $\theta_{i,j}^{n+1}$. The whole procedure is then repeated cyclically.

This approach to the solution of the (θ, ω, ψ) system is especially effective with stable implicit difference schemes, since they allow large time steps in the calculations.

4.2 CHARACTERISTICS OF COMPUTATION ALGORITHMS

We now consider some of the properties of the computational algorithms for the numerical analysis of convective equations based on the Boussinesque system (4.4). One characteristic that may be attributed to the approximate nature of the boundary conditions for the vorticity function ω was discussed at the end of the previous section, where the procedure for the numerical scheme for the system (4.4) was presented.

Numerical solution of the system requires the solution of the Poisson equation for the stream function

$$\frac{\partial^2 \psi}{\partial x^2} + \frac{\partial^2 \psi}{\partial y^2} = -\omega \qquad (4.11)$$

at every time level. A number of efficient numerical methods are available for such equations. The simplest and most efficient method is probably the Saidel method, together with successive overrelaxation. The equation for the difference solution may be written as

$$\psi_{i,j}^{s+1} = (1-\gamma)\psi_{i,j}^s + \frac{1}{2}\left[\frac{1}{(\Delta x)^2} + \frac{1}{(\Delta y)^2}\right]^{-1}$$
$$\cdot \left[\frac{\psi_{i+1,j}^s + \psi_{i-1,j}^{s+1}}{(\Delta x)^2} + \frac{\psi_{i,j+1}^s + \psi_{i,j-1}^{s+1}}{(\Delta y)^2} + \omega_{i,j}^{n+1}\right] \qquad (4.12)$$

where s is the iteration number, and γ is the relaxation parameter. The method appears to be useful, since it requires a minimum amount of computer storage and its optimal relaxation parameter

$$\gamma = 2\left[1 + \sqrt{1 - \left(\frac{\cos \pi \Delta x + \cos \pi \Delta y}{2}\right)^2}\right]^{-1} \qquad (4.13)$$

provides a sufficiently high speed of convergence for the iteration procedure. The value of $\psi_{i,j}^n$ at the previous time step is usually taken as the initial approximation. The iteration procedure is terminated when the condition

$$\max_{i,j} \left| \frac{\psi_{i+1,j}^{s+1} - 2\psi_{i,j}^{s+1} + \psi_{i-1,j}^{s+1}}{(\Delta x)^2} + \frac{\psi_{i,j+1}^{s+1} - 2\psi_{i,j}^{s+1} + \psi_{i,j-1}^{s+1}}{(\Delta y)^2} + \omega_{i,j}^{n+1} \right| \leq \epsilon_\psi$$
$$(4.14)$$

or

$$\frac{\max\limits_{i,j} |\psi_{i,j}^{s+1} - \psi_{i,j}^{s}|}{\max\limits_{i,j} |\psi_{i,j}^{s+1}|} \leq \epsilon_\psi \qquad (4.15)$$

is satisfied, thus $\psi_{i,j}^{s+1} = \psi_{i,j}^{n+1}$. A computation accuracy of $\epsilon_\psi \approx 10^{-3}$ usually requires about 10 or 20 iterations. As the solution approaches the steady-state condition, the number of necessary iterations is reduced to one to three.

Often, stream functions are evaluated by the iteration procedure based on the fractional-step method. The numerical algorithm in this case is divided into the two stages:

$$\frac{\psi_{i,j}^{s+1/2} - \psi_{i,j}^{s}}{0.5\tau'} = \frac{\psi_{i+1,j}^{s+1/2} - 2\psi_{i,j}^{s+1/2} + \psi_{i-1,j}^{s+1/2}}{(\Delta x)^2} + \frac{\psi_{i,j+1}^{s} - 2\psi_{i,j}^{s} + \psi_{i,j-1}^{s}}{(\Delta y)^2} + \omega_{i,j}^{n+1} \qquad (4.16)$$

$$\frac{\psi_{i,j}^{s+1} - \psi_{i,j}^{s+1/2}}{0.5\tau'} = \frac{\psi_{i+1,j}^{s+1/2} - 2\psi_{i,j}^{s+1/2} + \psi_{i-1,j}^{s+1/2}}{(\Delta x)^2} + \frac{\psi_{i,j+1}^{s+1} - 2\psi_{i,j}^{s+1} + \psi_{i,j-1}^{s+1}}{(\Delta y)^2} + \omega_{i,j}^{n+1}$$

where τ' is the iteration parameter. The implicit equations of system (4.16) are solved by the factorization method. The recurrence formulas for the factorization coefficients for the first equation of system (4.16), when $\Delta x = \Delta y = h$, are

$$\alpha_i = \left(2 + \frac{2h^2}{\tau} - \alpha_{i-1}\right)^{-1} \qquad \beta_i = (\beta_{i-1} + F^s)\alpha_i$$
$$F^s = \psi_{i,j+1}^{s} + \left(\frac{2h^2}{\tau} - 2\right)\psi_{i,j}^{s} + \psi_{i,j-1}^{s} + h^2\omega_{i,j}^{n+1} \qquad \alpha_0 = \beta_0 = 0 \qquad (4.17)$$

$\psi_{i,j}^{s+1/2}$ is calculated from

$$\psi_{i,j}^{s+1/2} = \alpha_i \psi_{i+1,j}^{s+1/2} + \beta_i \qquad \psi_{i,j}^{s+1/2} = 0$$

CONVECTIVE HEAT TRANSFER

For the second equation,

$$\alpha_j = \left(2 + \frac{2h^2}{\tau} - \alpha_{j-1}\right)^{-1} \qquad \beta_j = (\beta_{j-1} + F^{s+1/2})\alpha_j$$

$$F^{s+1/2} = \psi_{i+1,j}^{s+1/2} + \left(\frac{2h^2}{\tau} - 2\right)\psi_{i,j}^{s+1/2} + \psi_{i-1,j}^{s+1/2} + h^2 \psi_{i,j}^{n+1} \qquad \alpha_0 = \beta_0 = 0$$

(4.18)

$\psi_{i,j}^{s+1}$ is computed from the recurrence formula

$$\psi_{i,j}^{s+1} = \alpha_j \psi_{i,j+1}^{s+1} + \beta_j \qquad \psi_{i,J}^{s+1} = 0$$

The accuracy of iteration procedure (4.16) is controlled by condition (4.14) or (4.15).

It should be noted that the second method for solving system (4.11), though more time consuming, seems advantageous since its rate of convergence is higher than that of the Saidel method by a factor of 1.5.

For the numerical solution of convective heat-transfer problems, we usually assume some constant accuracy ϵ_ψ for the Poisson system (4.11). The value of ϵ_ψ is normally chosen in accordance with the accuracy desired for the functions ω and θ.

The solution of the unsteady $(\theta, \omega, \psi,)$ system is often used for steady-state problems. The steady solution is obtained as $t \to \infty$. Since the development of the steady solution is of no interest here, it is usually only required that the solution have some desired accuracy ϵ_ψ. The solution of Eq. (4.11) with such accuracy ϵ_ψ at each time level seems useful, however, since that improves the stability of the computation. Numerical experiments have demonstrated that a reasonable choice of ϵ_ψ leads to a decrease in the total computation time. The accuracy of system (4.11) should be based on the accuracy of the functions ω and θ at the time step considered. For example, we can choose

$$\epsilon_\psi{}^n = \min(\epsilon_\omega{}^n, \epsilon_\theta{}^n) \qquad (4.19)$$

where $\epsilon_\omega{}^n = \max_{i,j} |\omega_{i,j}^{n+1} - \omega_{i,j}^n| / \max_{i,j} |\omega_{i,j}^{n+1}|$
$\epsilon_\theta{}^n = \max_{i,j} |\theta_{i,j}^{n+1} - \theta_{i,j}^n| / \max_{i,j} |\theta_{i,j}^{n+1}|$

Numerical experiments have revealed that, in order to reach the desired accuracy at each time step, two to four iterations of the Poisson equation are necessary.

System (4.4) involves the velocity projections u and v on the x and y axes, respectively. They are defined in terms of the stream function ψ as $u = \partial\psi/\partial y$ and $v = -\partial\psi/\partial x$. In forming numerical algorithms, it is useful to calculate the values of $u_{i,j}$ and $v_{i,j}$ immediately after the $\psi_{i,j}^{n+1}$'s have been determined and to store them in the computer if possible.

The successful solution of many practical problems often requires not only some mean characteristics of the heat transfer from heated surfaces but also other details, such as local heat-flux distributions, structures of temperature fields, and characteristics of transient processes.

Heat fluxes are generally expressed in terms of the Nusselt number which, in a dimensionless form, is

$$\text{Nu} = -\int_\Gamma \frac{\partial \theta}{\partial n} dl \qquad (4.20)$$

where $\partial\theta/\partial n$ is the temperature gradient along the normal directed to the interior of the region. The integration is carried out over the part of the boundary for which the heat flux is being determined.

Heat fluxes into the system are usually assumed positive, and those out of the system negative. These fluxes are denoted as Nu^+ and Nu^-, respectively. The integration to determine Nu^+ and Nu^- should be carried out over the boundary intervals in which the derivative $\partial\theta/\partial n$ does not change sign.

For the actual computation of the integral in expression (4.20), Simpson's formula may be used in the form

$$\int_a^b \frac{\partial \theta}{\partial n} dl \approx \frac{b-a}{3K} \sum_{k=1(2)}^{K-1} (y_k + 4y_{k+1} + y_{k+2}) \qquad (4.21)$$

where $y_k = (\partial\theta/\partial n)_k$, and the summation is carried out with respect to index k from 1 to K (K is even) with a step of 2. The derivative $\partial\theta/\partial n$ may be evaluated using the three-point difference equation of the second order. At the boundary points coinciding with the x axis it can be written as

$$\left(\frac{\partial \theta}{\partial n}\right)_{i,j} = -\frac{3\theta_{i,0} - 4\theta_{i,1} + \theta_{i,2}}{2h}$$

With very large temperature gradients in the region adjacent to the boundary, the first-order two-point formula

$$\left(\frac{\partial \theta}{\partial n}\right)_{i,j} = \frac{\theta_{i,1} - \theta_{i,0}}{h} \qquad (4.22)$$

is recommended.

If there are heat fluxes of different signs at the boundary one may obtain Nu^+ and Nu^- by first calculating

$$q = \int_{\Gamma} \frac{\partial \theta}{\partial n} dl \quad \text{and} \quad |q| = \int_{\Gamma} \left|\frac{\partial \theta}{\partial n}\right| dl$$

Then

$$Nu^+ = \frac{1}{2}(|q| - q) \quad \text{and} \quad Nu^- = -\frac{1}{2}(|q| + q) \qquad (4.23)$$

Additional heat transfer from a heated boundary to a cold one (as compared to heat conduction) is one of the most important characteristics of convection and may be expressed in terms of the convective Nusselt number:

$$Nu^k = \frac{Nu(Gr, Pr)}{Nu(Gr = 0, Pr)} \qquad (4.24)$$

Difference schemes for the (θ, ω, ψ) system involving free-convection cases (Re = 1) and the extensions of these schemes to forced-convection problems will be discussed in the following sections.

4.3 NUMERICAL PROCEDURES FOR HEAT-CONVECTION PROCESSES WITH LOW AND MEDIUM RATES

The rates of heat-convection processes governed by system (4.4) depend on such parameters as the Grashof, Prandtl, and Reynolds numbers. The rate of free convection (Re = 1) and convective heat transfer may be expressed in terms of the Rayleigh number alone (Ra = Gr Pr). We now present finite-difference procedures for the convective heat transfer over the following ranges of the controlling parameter: $0.1 \leq \text{Pr} \leq 10^3$; $0 \leq \text{Gr} \leq 10^6$ (but $\text{Ra} \leq 10^6$); $1 \leq \text{Re} \leq 10^2$.

The original system (4.4) contains three equations, namely, those for temperature, vorticity, and stream function. The first two equations are parabolic; the Poisson equation for the stream function is elliptic. All the equations under consideration involve solution of the stream-function equation at every time (iteration) level.

The most widely used numerical procedures for system (4.4) were discussed in the previous section. These procedures differ from one another only in the temperature and vorticity equations used. It may easily be seen that these equations differ only in that the coefficient 1/Pr appears in the temperature equation, and the term $\text{Gr}(\partial\theta/\partial x)$ (at $\varphi = \pi/2$) is contained in the vorticity equation.

The finite-difference approximations of these equations would naturally differ by the same terms. It is therefore convenient to consider only the model equation

$$\frac{\partial \phi}{\partial t} = \frac{1}{\text{Pr}} \left(\frac{\partial^2 \phi}{\partial x^2} + \frac{\partial^2 \phi}{\partial y^2} \right) - u \frac{\partial \phi}{\partial x} - v \frac{\partial \phi}{\partial y} + f \qquad (4.25)$$

which for $\phi = \theta$ and $f = 0$ becomes the temperature equation, and for $\phi = \omega$, $\text{Pr} = 1$, and $f = \text{Gr}(\partial\theta/\partial x)$ becomes the vorticity equation.

The simplest (but not always the best) numerical procedures for the (θ, ω, ψ) system describing heat convection in fluids are those that employ explicit finite-difference schemes. Among these, the most commonly used is the scheme in which the diffusion and convective terms are approximated by symmetrical difference relations (central differences) as

$$\frac{\phi_{i,j}^{n+1} - \phi_{i,j}^n}{\tau} = \frac{1}{\text{Pr } h^2}(\phi_{i+1,j}^n + \phi_{i-1,j}^n + \phi_{i,j-1}^n + \phi_{i,j+1}^n - 4\phi_{i,j}^n)$$

$$- u_{i,j} \frac{\phi_{i+1,j}^n - \phi_{i-1,j}^n}{2h} - v_{i,j} \frac{\phi_{i,j+1}^n - \phi_{i,j-1}^n}{2h} + f_{i,j} \qquad (4.26)$$

where $u_{i,j} = (\psi_{i,j+1} - \psi_{i,j-1})/2h$
$v_{i,j} = (\psi_{i,j-1} - \psi_{i,j+1})/2h$

This scheme is of the order of approximation $O(\tau + h^2)$. The stability condition places a severe restriction on the time step,

CONVECTIVE HEAT TRANSFER

$$\tau \leqslant \frac{h^2}{2\mathrm{Pr}(2 + \delta\psi)} \qquad \delta\psi = \max_{i,j} (|\psi_{i+1,j} - \psi_{i-1,j}|, |\psi_{i,j+1} - \psi_{i,j-1}|) \quad (4.27)$$

resulting in a large increase in the computation time. Moreover, difference scheme (4.26) is nonconservative, resulting in an appreciable disturbance of the heat-flux balance at the boundaries of the computation region for strongly nonuniform temperatures. These disadvantages are compensated by the simplicity of the numerical algorithm, which is why the scheme is so popular.

The use of the integrointerpolation method for system (4.25) readily gives a conservative difference scheme of the same order of approximation as (4.26). For this purpose, (4.25) may be written in a divergence form as

$$\frac{\partial \phi}{\partial t} = \frac{\partial}{\partial x}\left(\frac{1}{\mathrm{Pr}}\frac{\partial \phi}{\partial x} - u\phi\right) + \frac{\partial}{\partial y}\left(\frac{1}{\mathrm{Pr}}\frac{\partial \phi}{\partial y} - v\phi\right) + f \quad (4.28)$$

and integrated over the mesh $D_i(x_{i-1/2} \leqslant x \leqslant x_{i+1/2}, y_{j-1/2} \leqslant y \leqslant y_{j+1/2})$. (See Fig. 6.)

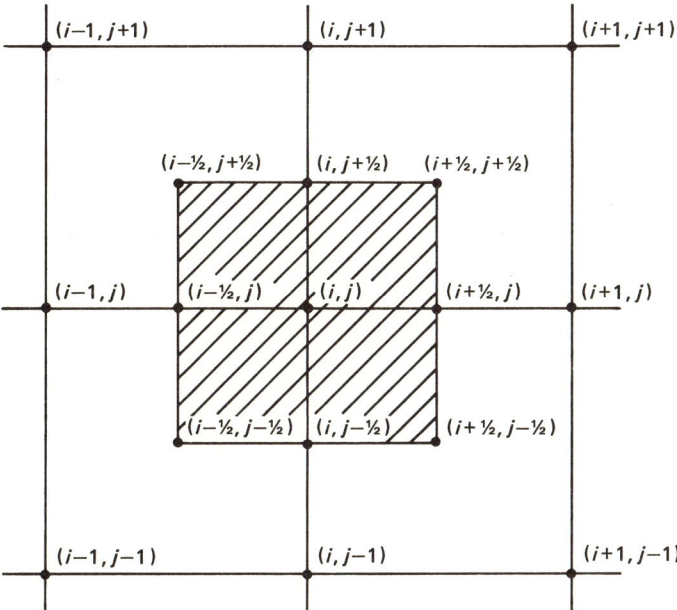

Fig. 6 The mesh D_i.

The use of the Green formula

$$\iint_{D_i} \left(\frac{\partial \phi_1}{\partial x} + \frac{\partial \phi_2}{\partial y}\right) dx\, dy = \oint_{\Gamma} (\phi_1\, dy - \phi_2\, dx) \qquad (4.29)$$

gives

$$\iint_{D_i} \left(\frac{\partial \phi}{\partial t} - f\right) dx\, dy = \oint_{\Gamma} \left(\frac{1}{\Pr}\frac{\partial \phi}{\partial x} - u\phi\right) dy - \oint_{\Gamma} \left(\frac{1}{\Pr}\frac{\partial \phi}{\partial y} - v\phi\right) dx \qquad (4.30)$$

Contour Γ bounding mesh D_i will be taken counterclockwise. It may easily be seen that the first integral on the right-hand side of (4.30) vanishes for horizontal sections of the boundary, and the second integral vanishes for vertical sections:

$$\iint_{D_i} \left(\frac{\partial \phi}{\partial t} - f\right) dx\, dy = \int_{y_{j-1/2}}^{y_{j+1/2}} \left(\frac{1}{\Pr}\frac{\partial \phi}{\partial x} - u\phi\right)_{i-1/2}^{i+1/2} dy$$

$$+ \int_{x_{i-1/2}}^{x_{i+1/2}} \left(\frac{1}{\Pr}\frac{\partial \phi}{\partial y} - v\phi\right)_{j-1/2}^{j+1/2} dx$$

$$= \frac{1}{\Pr}\left(\int_{y_{j-1/2}}^{y_{j+1/2}} \frac{\partial \phi}{\partial x}\bigg|_{i-1/2}^{i+1/2} dy + \int_{x_{i-1/2}}^{x_{i+1/2}} \frac{\partial \phi}{\partial y}\bigg|_{j-1/2}^{j+1/2} dx\right)$$

$$- \int_{y_{j-1/2}}^{y_{j+1/2}} u\phi \bigg|_{i-1/2}^{i+1/2} dy - \int_{x_{i-1/2}}^{x_{i+1/2}} v\phi \bigg|_{j-1/2}^{j+1/2} dx$$

Integration gives

$$h^2 \left(\frac{\partial \phi}{\partial t} - f\right)_{i,j} = \frac{1}{\Pr}(\phi_{i+1,j} + \phi_{i-1,j} + \phi_{i,j+1} + \phi_{i,j-1} - 4\phi_{i,j})$$

$$- h[(u_{i+1/2,j}\phi_{i+1/2,j} - u_{i-1/2,j}\phi_{i-1/2,j})$$

$$+ (v_{i,j+1/2}\phi_{i,j+1/2} - v_{i,j-1/2}\phi_{i,j-1/2})] \qquad (4.31)$$

CONVECTIVE HEAT TRANSFER

The first term on the right-hand side of (4.31) is diffusion, and the second is convection. If $\phi_{i\pm1/2,j} = 0.5(\phi_{i\pm1,j} + \phi_{i,j})$ and $\phi_{i,j\pm1/2} = 0.5(\phi_{i,j} + \phi_{i,j\pm1})$ are assumed for the convection terms, and the velocity projections u and v are expressed in terms of stream functions at the integral grid points as

$$u_{i+1/2,j} = \frac{\psi_{i+1/2,j+1/2} - \psi_{i+1/2,j-1/2}}{h}$$

$$= \frac{\psi_{i,j+1} + \psi_{i+1,j+1} - \psi_{i,j-1} - \psi_{i+1,j-1}}{4h}$$

$$u_{i-1/2,j} = \frac{\psi_{i-1/2,j+1/2} - \psi_{i-1/2,j-1/2}}{h}$$

$$= \frac{\psi_{i-1,j+1} + \psi_{i,j+1} - \psi_{i,j-1} - \psi_{i+1,j-1}}{4h}$$

(4.32)

$$v_{i,j+1/2} = -\frac{\psi_{i+1/2,j+1/2} - \psi_{i-1/2,j+1/2}}{h}$$

$$= -\frac{\psi_{i+1,j+1} + \psi_{i+1,j} - \psi_{i-1,j} - \psi_{i-1,j+1}}{4h}$$

$$v_{i,j-1/2} = -\frac{\psi_{i+1/2,j-1/2} - \psi_{i-1/2,j-1/2}}{h}$$

$$= -\frac{\psi_{i+1,j} + \psi_{i+1,j-1} - \psi_{i-1,j} - \psi_{i-1,j-1}}{4h}$$

then the explicit conservative difference scheme

$$\frac{h^2}{\tau}(\phi_{i,j}^{n+1} - \phi_{i,j}^n) = \frac{1}{\Pr}(\phi_{i+1,j}^n + \phi_{i-1,j}^n + \phi_{i,j+1}^n + \phi_{i,j-1}^n - 4\phi_{i,j}^n)$$

$$- h[u_{i+1/2,j}(\phi_{i+1,j}^n + \phi_{i,j}^n) - u_{i-1/2,j}(\phi_{i-1,j}^n + \phi_{i,j}^n)$$

$$+ v_{i,j+1/2}(\phi_{i,j+1}^n + \phi_{i,j}^n) - v_{i,j-1/2}(\phi_{i,j-1}^n + \phi_{i,j}^n)] + h^2 f_{i,j}$$

(4.33)

may easily be obtained.

Scheme (4.33) approximates differential equation (4.25) with order $O(\tau + h^2)$. It is stable at

$$\tau \leqslant \min\left[\frac{h^2}{4\Pr}, \frac{h}{\max\limits_{i,j}(|u_{i,j}|, |v_{i,j}|)}\right]$$

(4.34)

Though the order of approximation and stability condition for difference schemes (4.26) and (4.27) are almost the same, the latter is advantageous since it describes the heat balance more accurately.

Because of the simplicity of the algorithms of these two explicit difference schemes, computer programs may readily be developed for the desired numerical solution. However, when a systematic numerical analysis of a class of problems is required, implicit procedures seem more effective. The latter, though requiring more time for programming, considerably reduce the required computer time.

Among the implicit schemes, the most widely used is the horizontal-vertical scheme of the form

$$\phi_{i,j}^{n+1/2} = \phi_{i,j}^n + \frac{\tau}{2}(\delta_x \phi^{n+1/2} + \delta_y \phi^n + f_{i,j}^{n+1/2})$$

$$\phi_{i,j}^{n+1} = \phi_{i,j}^{n+1/2} + \frac{\tau}{2}(\delta_x \phi^{n+1/2} + \delta_y \phi^{n+1} + f_{i,j}^{n+1/2})$$

(4.35)

Here, the difference operators

$$\delta_x \phi = \frac{1}{\Pr h^2}(\phi_{i+1,j} - 2\phi_{i,j} + \phi_{i-1,j}) - u_{i,j}\frac{\phi_{i,j+1,j} - \phi_{i-1,j}}{2h}$$

$$\delta_y \phi = \frac{1}{\Pr h^2}(\phi_{i,j+1} - 2\phi_{i,j} + \phi_{i,j-1}) - v_{i,j}\frac{\phi_{i,j+1} - \phi_{i,j-1}}{2h}$$

(4.36)

approximate, respectively, the differential operators

$$L_x \phi = \frac{1}{\Pr}\frac{\partial^2 \phi}{\partial x^2} - u\frac{\partial \phi}{\partial x} \quad L_y \phi = \frac{1}{\Pr}\frac{\partial^2 \phi}{\partial y^2} - v\frac{\partial \phi}{\partial y}$$

with error $O(h^2)$.

To calculate $\phi_{i,j}^{n+1}$ with $\phi_{i,j}^n$ known, the first and second equations of (4.35) are successively solved by the factorization method. First, with fixed values of index j ($j = 1, 2, 3, \ldots, J-1$), the values of $\phi_{1,j}^{n+1/2}$, $\phi_{2,j}^{n+1/2}$, $\phi_{3,j}^{n+1/2}$, ... are successively determined (horizontal factorization). After all $\phi_{i,j}^{n+1/2}$'s have been found, $\phi_{i,1}^{n+1}$, $\phi_{i,2}^{n+1}$, $\phi_{i,3}^{n+1}$, ..., $\phi_{i,J-1}^{n+1}$ are determined in turn from the second equation (vertical factorization) for fixed i ($i = 1, 2, 3, \ldots, J-1$).

Scheme (4.35) approximates the original differential equations with

CONVECTIVE HEAT TRANSFER

accuracy $O(\tau + h^2)$. The stability analysis of the scheme is extremely complicated because of its nonlinearity. Extensive numerical experiments demonstrate that the computation remains stable up to $\tau \approx h$.

Similarly to scheme (4.33), difference scheme (4.35) may be expressed in a divergence form. The general idea of the algorithm then remains the same, but the form of the difference operators changes to

$$\delta_x \phi = \frac{1}{\Pr h^2} (\phi_{i+1,j} - 2\phi_{i,j} + \phi_{i-1,j})$$

$$- \frac{u_{i+1/2,j}(\phi_{i,j} + \phi_{i+1,j}) - u_{i-1/2,j}(\phi_{i,j} + \phi_{i-1,j})}{h}$$

(4.37)

$$\delta_y \phi = \frac{1}{\Pr h^2} (\phi_{i,j+1} - 2\phi_{i,j} + \phi_{i,j-1})$$

$$- \frac{v_{i,j+1/2}(\phi_{i,j} + \phi_{i,j+1}) - v_{i,j-1/2}(\phi_{i,j} + \phi_{i,j-1})}{h}$$

The velocity projections $u_{i \pm 1/2, j}$ and $v_{i, j \pm 1/2}$ are defined by Eqs. (4.32).

Computational experience has revealed that for convective heat transfer in regions with pronounced nonuniformity of temperature, approximation (4.37) is advantageous, since it gives better balances of heat fluxes.

Implicit difference schemes of the predictor-corrector type and stabilization schemes seem very promising for unsteady-state convective problems. These schemes may also be extended to the case of a system of convective heat-transfer equations using the horizontal-vertical scheme.

A general procerure for solving convective heat-transfer problems by numerical algorithms based on the numerical solution of (θ, ω, ψ) system will now be considered in detail.

The procedure for the solution of the (θ, ω, ψ) system in the case of steady-state convective problems is essentially the same as that presented in Sec. 4.1. The steady-state solution is obtained at $t \to \infty$. The solution of the (θ, ω, ψ) system at the $(n + 1)$th level is found in the following way. First, the $\omega_{i,j}^{n+1}$'s are evaluated at points near the boundary of the difference grid, from the difference equation for the stream function $\nabla^2 \psi = -\omega$. Then, with the resultant $\omega_{i,j}^{n+1}$ taken as the boundary values, the $\omega_{i,j}^{n+1}$'s in the remainder of the region are found from the analog of the vorticity equation. Thereafter, with the assumption that the $\psi_{i,j}^n$'s are given at the primary boundary, the $\psi_{i,j}^{n+1}$ are evaluated inside the computation region from (4.12) or (4.16) by iteration. The boundary values $\psi_{i,j}^{n+1}$ are now corrected to satisfy the boundary condition for the normal derivative $\partial \psi/\partial n$ (for the no-slip condition, $\partial \psi/\partial n = 0$). Finally, $\theta_{i,j}^{n+1}$ is found from the temperature equation.

For unsteady-state cases, the numerical procedure for the (θ, ω, ψ) system is somewhat more complicated. In order that the approximate solutions for any time t^n be of the desired accuracy (not only for $t \to \infty$ as in steady-state cases), iteration procedures are required at every time level to determine accurate $\theta_{i,j}^{n+1}$, $\omega_{i,j}^{n+1}$, and $\psi_{i,j}^{n+1}$.

The iteration procedure is carried out as follows. First, $\theta_{i,j}^{n+1}$, $\omega_{i,j}^{n+1}$, and $\psi_{i,j}^{n+1}$ are evaluated in the usual way. The evaluation is then repeated with only one difference: In the computation of the velocity components $u_{i,j}$ and $v_{i,j}$ and source $\text{Gr}(\partial \theta/\partial x)$ in the vorticity equation, $\psi_{i,j}^{n+1}$ and $\theta_{i,j}^{n+1}$ are substituted for $\psi_{i,j}^n$ and $\theta_{i,j}^n$. The new results for $\theta_{i,j}^{n+1}$, $\omega_{i,j}^{n+1}$, and $\psi_{i,j}^{n+1}$ are then compared with the old values. If the required accuracy has not been achieved, the computation is repeated. Now, for the calculation of $u_{i,j}$, $v_{i,j}$, and $\text{Gr}(\partial \theta/\partial x)$, more accurate values of $\theta_{i,j}^{n+1}$, $\omega_{i,j}^{n+1}$, $\psi_{i,j}^{n+1}$, etc., are used. Once the required accuracy is achieved, the computation of θ, ω, and ψ for subsequent times t^{n+2}, t^{n+3}, ... is started.

The difference schemes considered here so far have an error of $O(h^2)$, thus giving a satisfactory accuracy even with grids of $h \approx 0.05$. The range of validity of the schemes is, however, restricted, owing to their nonmonotonic nature; i.e., they do not satisfy the maximum principle. The monotonicity condition that follows from the requirement of positive coefficients is of the form

$$h < \frac{2}{\max_{i,j}(|u_{i,j}|, |v_{i,j}|)} \tag{4.38}$$

for the difference schemes discussed above, which is practically impossible at $\text{Ra} > 10^6$.

For a fixed h, therefore, the computation procedure leads to oscillations in temperature, vorticity, and stream function about their mean values. Decreasing the time step τ does not improve the situation.

4.4 HIGH-RATE CONVECTION PROCESSES

The high temperatures and pressures that may be encountered in many practical problems lead to high-rate convective processes that require special numerical procedures for their solution. One of the main difficulties in this respect is the need to satisfy the competing requirements of stability and accuracy: Schemes may be stable, but the approximated viscosity effect is large and may exceed the physical one; or they may be accurate, with the approximated viscosity effect being small, but the computation process is slowly convergent or does not converge at all.

The use of the finite-difference procedures described in the previous section and based on central differencing of convective terms involves, for the present situation, certain mathematical difficulties resulting from restrictions on the space steps of the grid due to the monotonicity requirement (4.38).

These difficulties naturally disappear in the one-sided approximation of convective terms. The difference procedure often used for convective problems, in which the convective terms are approximated by one-sided differences and which accounts for the velocity-vector direction ("up the flow"), is

$$\frac{\phi_{i,j}^{n+1} - \phi_{i,j}^n}{\tau} = \frac{1}{\Pr h^2}(\phi_{i+1,j}^n + \phi_{i-1,j}^n + \phi_{i,j-1}^n + \phi_{i,j+1}^n - 4\phi_{i,j}^n)$$

$$- \left(\frac{u_{i,j} - |u_{i,j}|}{2} \frac{\phi_{i+1,j}^n - \phi_{i,j}^n}{h} + \frac{u_{i,j} + |u_{i,j}|}{2} \frac{\phi_{i,j}^n - \phi_{i-1,j}^n}{h} \right)$$

$$- \left(\frac{v_{i,j} - |v_{i,j}|}{2} \frac{\phi_{i,j+1}^n - \phi_{i,j}^n}{h} + \frac{v_{i,j} + |v_{i,j}|}{2} \frac{\phi_{i,j}^n - \phi_{i,j-1}^n}{h} \right) + f_{i,j}$$

(4.39)

where $u_{i,j} = (\psi_{i,j+1} - \psi_{i,j-1})/(2h)$
$v_{i,j} = -(\psi_{i,j+1} - \psi_{i,j-1})/(2h)$

It may easily be verified that this difference scheme is really monotonic (the coefficients at $\phi_{i+1,j}$, $\phi_{i-1,j}$, $\phi_{i,j+1}$, and $\phi_{i,j-1}$ are positive, and the modulus of their sum does not exceed that of the coefficient of $\phi_{i,j}$). The stability condition is

$$\tau \leqslant \min_{i,j} \left(\frac{h^2}{4\Pr}, \frac{h}{|u_{i,j}| + |v_{i,j}|} \right) \qquad (4.40)$$

Numerical experiments reveal that with a suitable computation procedure this scheme gives monotonic solutions up to $\text{Ra} \approx 10^{10}$. However, the scheme has an important disadvantage: It has first-order accuracy not only with respect to τ but also with respect to h, and therefore satisfactory accuracy requires a very small grid spacing.

To overcome this, difference schemes of a higher order of accuracy are desirable. In particular, the following finite-difference scheme based on the method developed by Samarsky [6] may be recommended:

$$\frac{\phi_{i,j}^{n+1} - \phi_{i,j}^n}{\tau} = \frac{\phi_{i+1,j}^n - 2\phi_{i,j}^n + \phi_{i-1,j}^n}{(1 + 0.5|u_{i,j}|\,\mathrm{Pr}\,h)\mathrm{Pr}\,h^2} + \frac{\phi_{i,j+1}^n - 2\phi_{i,j}^n + \phi_{i,j-1}^n}{(1 + 0.5|v_{i,j}|\,\mathrm{Pr}\,h)\mathrm{Pr}\,h^2}$$

$$- \frac{u_{i,j} - |u_{i,j}|}{2} \frac{\phi_{i+1,j}^n - \phi_{i,j}^n}{h} + \frac{u_{i,j} + |u_{i,j}|}{2} \frac{\phi_{i,j}^n - \phi_{i-1,j}^n}{h}$$

$$- \frac{v_{i,j} - |v_{i,j}|}{2} \frac{\phi_{i,j+1}^n - \phi_{i,j}^n}{h} + \frac{v_{i,j} + |v_{i,j}|}{2} \frac{\phi_{i,j}^n - \phi_{i,j-1}^n}{h} + f_{i,j}$$

(4.41)

This scheme is not very much more complicated than (4.39), but it is monotonic and approximates the initial differential equation (4.25) with order $O(\tau + h^2)$. The stability condition is almost the same as (4.40).

Both schemes are nonconservative, and at large u and v the heat-flux balance is appreciably violated.

Let us construct a monotonically conservative scheme of the second order. To this end, the differential equation (4.28) is integrated over mesh D_i. The integration, described in the previous section, results in expression (4.31):

$$h^2 \left(\frac{\partial \phi}{\partial t} - f\right)_{i,j} = \frac{1}{\mathrm{Pr}} (\phi_{i+1,j} + \phi_{i-1,j} + \phi_{i,j+1} + \phi_{i,j-1} - 4\phi_{i,j})$$

$$- h[(u_{i+1/2,j}\phi_{i+1/2,j} - u_{i-1/2,j}\phi_{i-1/2,j})$$

$$+ (v_{i,j+1/2}\phi_{i,j+1/2} - v_{i,j-1/2}\phi_{i,j-1/2})]$$

For the difference scheme to be both conservative and monotonic, the conditions should be formulated to exclude the violation of monotonicity at any u and v. The conditions obviously will be determined from the velocity vector, i.e., the signs of u and v. Indeed, if the substitution

$$u_{i+1/2,j}\phi_{i+1/2,j} - u_{i-1/2,j}\phi_{i-1/2,j} = \begin{cases} u_{i+1/2,j}\phi_{i,j} - u_{i-1/2,j}\phi_{i-1,j} \\ \qquad\qquad\qquad \text{for } u > 0 \\ u_{i+1/2,j}\phi_{i+1,j} - u_{i-1/2,j}\phi_{i,j} \\ \qquad\qquad\qquad \text{for } u < 0 \end{cases}$$

(4.42)

CONVECTIVE HEAT TRANSFER

$$v_{i,j+1/2}\phi_{i,j+1/2} - v_{i,j-1/2}\phi_{i,j-1/2} = \begin{cases} v_{i,j+1/2}\phi_{i,j} - v_{i,j-1/2}\phi_{i,j-1} & \text{for } v > 0 \\ v_{i,j+1/2}\phi_{i,j+1} - v_{i,j-1/2}\phi_{i,j} & \text{for } v < 0 \end{cases}$$

(4.42) cont.

is made in the difference scheme, then a monotonic and conservative difference scheme having the first order of approximation with respect to h is obtained.

The approximation order of the scheme may be increased by use of a conventional method. A Taylor-series expansion of the function ϕ about the point (x_i, y_j) gives, for the main terms of the error, the expression

$$R = \frac{h}{2}\left[\frac{\partial}{\partial x}\left(|u|\frac{\partial \phi}{\partial x}\right) + \frac{\partial}{\partial y}\left(|v|\frac{\partial \phi}{\partial y}\right)\right] + O(h^2)$$

The terms with opposite signs are now added into the difference scheme, with the conservativeness and monotonicity of the scheme not being violated. The result is a monotonic and conservative finite-difference scheme with order of approximation $O(h^2)$. With intermediate operations omitted, the final result is

$$\phi_{i,j}^{n+1} = \phi_{i,j}^n + \tau(\bar{\delta}_x \phi^n + \bar{\delta}_y \phi^n + f_{i,j}) \tag{4.43}$$

where

$$\bar{\delta}_x \phi = \frac{\phi_{i+1,j} - \phi_{i,j}}{(1 + 0.5|u_{i+1/2,j}|\Pr h)\Pr h^2} - \frac{\phi_{i,j} - \phi_{i-1,j}}{(1 + 0.5|u_{i-1/2,j}|\Pr h)\Pr h^2}$$

$$- \frac{u_{i+1/2,j} + |u_{i+1/2,j}|}{2h}\phi_{i,j} + \frac{u_{i-1/2,j} + |u_{i-1/2,j}|}{2h}\phi_{i-1,j}$$

$$- \frac{u_{i+1/2,j} - |u_{i+1/2,j}|}{2h}\phi_{i+1,j} + \frac{u_{i-1/2,j} - |u_{i-1/2,j}|}{2h}\phi_{i,j}$$

(4.44)

$$\bar{\delta}_y \phi = \frac{\phi_{i,j+1} - \phi_{i,j}}{(1 + 0.5|v_{i,j+1/2}| \Pr h) \Pr h^2} - \frac{\phi_{i,j} - \phi_{i,j-1}}{(1 + 0.5|v_{i,j-1/2}| \Pr h) \Pr h^2}$$

$$- \frac{v_{i,j+1/2} + |v_{i,j+1/2}|}{2h} \phi_{i,j} + \frac{v_{i,j-1/2} + |v_{i,j-1/2}|}{2h} \phi_{i,j-1}$$

$$- \frac{v_{i,j+1/2} - |v_{i,j+1/2}|}{2h} \phi_{i,j+1} + \frac{v_{i,j-1/2} - |v_{i,j-1/2}|}{2h} \phi_{i,j}$$

(4.44)
cont.

The coefficients $u_{i+1/2,j}$, $u_{i-1/2,j}$, $v_{i,j+1/2}$, and $v_{i,j-1/2}$ are calculated from formulas (4.32). It may easily be noted that if we assume $u_{i+1/2,j} = u_{i-1/2,j} = u_{i,j}$ and $v_{i,j+1/2} = v_{i,j-1/2} = v_{i,j}$, then the result is monotonic scheme (4.41).

Finite-difference schemes (4.39), (4.41), and (4.43) are explicit, and when they are used, a strong limitation must be imposed on the time step because of stability condition (4.40).

Implicit difference schemes are, in a sense, free of this disadvantage. They may therefore be more useful for convective heat-transfer problems. Let us present here an economic, monotonic, and conservative difference scheme of the second order based on the fractional-step method. This scheme has been used successfully for unsteady convective heat-transfer problems [16, 17, 18] and has proved to be very useful. The scheme is based on a method similar to that used for scheme (4.43) and may be expressed as

$$\phi_{i,j}^{n+1/2} = \phi_{i,j}^n + 0.5\tau(\bar{\delta}_x \phi^{n+1/2} + \bar{\delta}_y \phi^n + f_{i,j}^{n+1/2})$$
$$\phi_{i,j}^{n+1} = \phi_{i,j}^{n+1/2} + 0.5\tau(\bar{\delta}_x \phi^{n+1/2} + \bar{\delta}_y \phi^{n+1} + f_{i,j}^{n+1/2})$$
(4.45)

Difference operators $\bar{\delta}_x \phi$ and $\bar{\delta}_y \phi$ may be found from expressions (4.44). For the computation of $\phi_{i,j}^{n+1}$, the first equation of system (4.45) is solved; then, with $\phi_{i,j}^n$ known, the intermediate values of $\phi_{i,j}^{n+1/2}$ are found. The factorization method is used for solving this equation. The second of Eqs. (4.45) is then solved for $\phi_{i,j}^{n+1}$ in the same way. The factorization coefficients may be calculated from difference scheme (4.45) written in the form

$$a_x \phi_{i-1,j}^{n+1/2} - \left(c_x + \frac{2}{\tau}\right) \phi_{i,j}^{n+1/2} + b_x \phi_{i+1,j}^{n+1/2} = -F_x$$

$$a_y \phi_{i,j-1}^{n+1} - \left(c_y + \frac{2}{\tau}\right) \phi_{i,j}^{n+1} + b_y \phi_{i,j+1}^{n+1} = -F_y$$
(4.46)

where

$$a_x = [(1 + 0.5|u_{i-1/2,j}| \Pr h) \Pr h^2]^{-1} + \frac{u_{i-1/2,j} + |u_{i-1/2,j}|}{2h}$$

$$b_x = [(1 + 0.5|u_{i+1/2,j}| \Pr h) \Pr h^2]^{-1} - \frac{u_{i+1/2,j} - |u_{i-1/2,j}|}{2h}$$

$$c_x = a_x + b_x + \frac{1}{h}(u_{i+1/2,j} - u_{i-1/2,j})$$

$$F_x = a_y \phi^n_{i,j-1} - \left(c_y - \frac{2}{\tau}\right)\phi^n_{i,j} + b_y \phi^n_{i,j+1} + f^{n+1/2}_{i,j}$$

$$a_y = [(1 + 0.5|v_{i,j-1/2}| \Pr h) \Pr h^2]^{-1} + \frac{v_{i,j-1/2} + |v_{i,j-1/2}|}{2h}$$

$$b_y = [(1 + 0.5|v_{i,j+1/2}| \Pr h) \Pr h^2]^{-1} - \frac{v_{i,j+1/2} - |v_{i,j-1/2}|}{2h}$$

$$c_y = a_y + b_y + \frac{1}{h}(v_{i,j+1/2} - v_{i,j-1/2})$$

$$F_y = a_x \phi^{n+1/2}_{i-1,j} - \left(c_x - \frac{2}{\tau}\right)\phi^{n+1/2}_{i,j} + b_x \phi^{n+1/2}_{i+1,j} + f^{n+1/2}_{i,j}$$

As in scheme (4.43), $u_{i-1/2,j}$, $u_{i+1/2,j}$, $v_{i,j+1/2}$, and $v_{i,j-1/2}$ may be calculated from (4.32), where

$$u_{i+1/2,j} = \frac{1}{2}(u^{n+1}_{i+1/2,j} + u^n_{i+1/2,j}) \qquad u_{i-1/2,j} = \frac{1}{2}(u^{n+1}_{i-1/2,j} + u^n_{i-1/2,j})$$

$$v_{i,j+1/2} = \frac{1}{2}(v^{n+1}_{i,j+1/2} + v^n_{i,j+1/2}) \qquad v_{i,j-1/2} = \frac{1}{2}(v^{n+1}_{i,j-1/2} + v^n_{i,j-1/2})$$

The recurrence formulas for calculating the factorization coefficients required in the first of Eqs. (4.45) are

$$\alpha_i = b_x \left(c_x + \frac{2}{\tau} - a_x \alpha_{i-1}\right)^{-1} \qquad \beta_i = (a_x \beta_{i-1} + F_x) \frac{\alpha_i}{b_x}$$

For the second equation,

$$\alpha_j = b_y \left(c_y + \frac{2}{\tau} - a_y \alpha_{j-1}\right)^{-1} \qquad \beta_j = (a_y \beta_{j-1} + F_y) \frac{\alpha_j}{b_y}$$

Numerical experiments reveal that difference scheme (4.45), with order of approximation $O(\tau + h^2)$, is stable if

$$\frac{2}{\tau} > \max_{i,j} \left(\frac{u_{i-1/2,j} - u_{i+1/2,j}}{h}, \frac{v_{i,j-1/2} - v_{i,j+1/2}}{h} \right) \quad (4.47)$$

and results in monotonic solutions up to $Ra \approx 10^{10}$. This difference scheme requires three to five times less computer time than the explicit scheme (4.43), other things being equal.

In the numerical analysis of convective heat-transfer problems using the finite-difference schemes just considered, the computational procedure for (θ, ω, ψ) systems is exactly the same as that using the difference schemes described in the previous section.

4.5 NUMERICAL ANALYSIS OF STEADY-STATE CONVECTIVE PROBLEMS

The development method is very useful for the numerical solution of steady-state convective heat-transfer problems. The solution of a steady-state problem is taken here as the limit that the solution of the unsteady problem approaches as $t \to \infty$. In the solution of steady-state problems by the development method, the approach to the steady state is of no practical interest; and in most cases it depends only on the number of successive approximations and has no basic physical meaning.

This allows the difference schemes developed here for the unsteady (θ, ω, ψ) system to be used for steady problems with τ taken as the iteration parameter. The computational algorithm is considerably simplified since exact solutions of the Poisson equations and iteration procedures for $\theta_{i,j}^{n+1}$, $\omega_{i,j}^{n+1}$, and $\psi_{i,j}^{n+1}$ are not required at any iteration level.

In a number of cases, the numerical analysis of steady-state convective problems may be based on a system of heat-convection equations of the form

$$\frac{\partial}{\partial x}\left(\frac{1}{Pr}\frac{\partial \theta}{\partial x} - u\theta\right) + \frac{\partial}{\partial y}\left(\frac{1}{Pr}\frac{\partial \theta}{\partial y} - v\theta\right) = 0$$

$$\frac{\partial}{\partial y}\left(\frac{\partial \omega}{\partial x} - u\omega\right) + \frac{\partial}{\partial y}\left(\frac{\partial \omega}{\partial y} - v\omega\right) = Gr\frac{\partial \theta}{\partial x} \quad (4.48)$$

$$\frac{\partial^2 \psi}{\partial x^2} + \frac{\partial^2 \psi}{\partial y^2} = -\omega$$

The numerical solution of this system may be obtained by solving its difference system with some available iteration procedure. Here, as in the case of unsteady problems, there are many finite-difference schemes and methods of solution. The difficulties involved are of the same nature as in unsteady cases.

The use of central differences for the approximation of convective terms leads to a nonmonotonic difference scheme that is valid only for $Ra < 10^6$. The study of high-rate convective heat-transfer processes requires difference schemes which are monotonic; their conservative properties are of great importance.

Consider the second-order conservative, monotonic finite-difference scheme approximating system (4.48). The scheme has been used for steady convective problems involving wide ranges of process parameters and has given good results. It employs the integrointerpolation method. A system of difference equations is obtained by integrating the original system (4.48) over a mesh $D_i(x_{i-1/2} \leqslant x \leqslant x_{i+1/2}, y_{j-1/2} \leqslant y \leqslant y_{j+1/2})$. The integration is carried out in the same way as in the case of difference scheme (4.43). This results in the difference analog for the temperature equation

$$\bar{\delta}_x \theta + \bar{\delta}_y \theta = 0 \tag{4.49}$$

where $\bar{\delta}_x \theta$ and $\bar{\delta}_y \theta$ are defined by (4.47). Solution of Eq. (4.49) for $\theta_{i,j}$ gives

$$\theta_{i,j} = \frac{a_\theta \theta_{i+1,j} + b_\theta \theta_{i-1,j} + c_\theta \theta_{i,j+1} + d_\theta \theta_{i,j-1}}{A_\theta} \tag{4.50}$$

where

$a_\theta = [(1 + 0.5|u_{i+1/2,j}| \Pr h) \Pr h]^{-1} - 0.5(u_{i+1/2,j} - |u_{i+1/2,j}|)$

$b_\theta = [(1 + 0.5|u_{i-1/2,j}| \Pr h) \Pr h]^{-1} + 0.5(u_{i-1/2,j} + |u_{i-1/2,j}|)$

$c_\theta = [(1 + 0.5|v_{i,j+1/2}| \Pr h) \Pr h]^{-1} - 0.5(v_{i,j+1/2} - |v_{i,j+1/2}|)$

$d_\theta = [(1 + 0.5|v_{i,j-1/2}| \Pr h) \Pr h]^{-1} + 0.5(v_{i,j-1/2} + |v_{i,j-1/2}|)$

$A_\theta = a_\theta + b_\theta + c_\theta + d_\theta$

The velocity projections u and v are calculated from Eqs. (4.32).

The same operations applied to the vorticity and stream-function equations give

$$\omega_{i,j} = \frac{a_\omega \omega_{i+1,j} + b_\omega \omega_{i-1,j} + c_\omega \omega_{i,j+1} + d_\omega \omega_{i,j-1}}{A_\omega}$$

$$+ \frac{(\text{Gr } h/2)(\theta_{i+1,j} - \theta_{i-1,j})}{A_\omega} \quad (4.51)$$

$$\psi_{i,j} = \frac{\psi_{i+1,j} + \psi_{i-1,j} + \psi_{i,j-1} + \psi_{i,j+1} + h^2 \omega_{i,j}}{4}$$

where

$$a_\omega = [(1 + 0.5|u_{i+1/2,j}|h)h]^{-1} - 0.5(u_{i+1/2,j} - |u_{i+1/2,j}|)$$
$$b_\omega = [(1 + 0.5|u_{i-1/2,j}|h)h]^{-1} + 0.5(u_{i-1/2,j} + |u_{i-1/2,j}|)$$
$$c_\omega = [(1 + 0.5|v_{i,j+1/2}|h)h]^{-1} - 0.5(v_{i,j+1/2} - |v_{i,j+1/2}|)$$
$$d_\omega = [(1 + 0.5|v_{i,j-1/2}|h)h]^{-1} + 0.5(v_{i,j-1/2} + |v_{i,j-1/2}|)$$
$$A_\omega = a_\omega + b_\omega + c_\omega + d_\omega$$

The iteration procedure for $\theta_{i,j}$, $\omega_{i,j}$, and $\psi_{i,j}$ is as follows:

$$\theta_{i,j}^{s+1} = (1 - \gamma_\theta)\theta_{i,j}^s + \frac{\gamma_\theta}{A_\theta}(a_\theta \theta_{i+1,j}^s + b_\theta \theta_{i-1,j}^{s+1} + c_\theta \theta_{i,j+1}^s + d_\theta \theta_{i,j-1}^{s+1})$$

$$\omega_{i,j}^{s+1} = (1 - \gamma_\omega)\omega_{i,j}^s + \frac{\gamma_\omega}{A_\omega}[a_\omega \omega_{i+1,j}^s + b_\omega \omega_{i-1,j}^{s+1} + c_\omega \omega_{i,j+1}^s$$
$$+ d_\omega \omega_{i,j-1}^{s+1} + 0.5 \text{ Gr } h(\theta_{i+1,j}^{s+1} - \theta_{i-1,j}^{s+1})]$$

$$\psi_{i,j}^{s+1} = (1 - \gamma_\psi)\psi_{i,j}^{s+1} + \frac{\gamma_\psi}{4}(\psi_{i+1,j}^s + \psi_{i-1,j}^{s+1} + \psi_{i,j+1}^s + \psi_{i,j-1}^{s+1} + h^2 \omega_{i,j}^{s+1})$$

(4.52)

where s is the iteration number, and γ_θ, γ_ω, and γ_ψ are the relaxation parameters.

The computational algorithm is arranged as follows. First, $\theta_{i,j}^{s+1}$'s are evaluated for the $(s + 1)$th iteration level; then the $\omega_{i,j}^{s+1}$'s are found; and finally $\psi_{i,j}^{s+1}$ is calculated.

The relaxation parameters γ_θ, γ_ω, and γ_ψ are chosen from numerical experiments. For convective heat-transfer problems with Rayleigh numbers Ra $\leq 10^5$, one may use γ_θ, $\gamma_\omega \approx 1$, and $\gamma_\psi \approx 1.6$. For higher Ra, steady solutions require considerably decreased γ_ω and increased γ_ψ. Thus, for

Ra $\approx 10^8$, one would use $\gamma_\theta \approx 1$, $\gamma_\omega \approx 0.01$, and $\gamma_\psi \approx 1.95$ (but less than 2!). Variation of the relaxation parameters may not only stabilize the computation procedure, but also increase considerably the rate of convergence of the iterations.

Two advantages of steady finite-difference schemes are their minimal computer storage requirements and their simplicity.

4.6 CONVECTIVE HEAT TRANSFER IN COMPRESSIBLE MEDIA

The numerical methods so far considered have been derived from the Boussinesque equations. This approach to the study of convective heat-transfer problems is very reasonable for many practical cases, and the results have been verified by numerous experiments. However, when the viscosity and thermal conductivity are strongly dependent on temperature, when the gas densities change appreciably with large temperature gradients (or, in accordance with the hydrostatic equilibrium conditions, with large compressibility parameters), then numerical investigations should be based on the complete system of compressible viscous gas equations.

The system of two-dimensional unsteady equations that governs the flow and heat transfer of a homogeneous, compressible, viscous gas is composed of the following momentum, continuity, and energy differential equations which, when written in terms of Cartesian coordinates in a dimensionless form as in [20], are

$$\frac{\partial u}{\partial t} + u\frac{\partial u}{\partial x} + v\frac{\partial u}{\partial y} = -\frac{1}{\kappa \rho C_\mu^2}\frac{\partial P}{\partial x} + \frac{1}{\rho C_R}\left(\frac{4}{3}\frac{\partial}{\partial x}\mu\frac{\partial u}{\partial x} + \frac{\partial}{\partial y}\mu\frac{\partial u}{\partial y} + \frac{\partial}{\partial y}\mu\frac{\partial v}{\partial x}\right.$$
$$\left. -\frac{2}{3}\frac{\partial}{\partial x}\mu\frac{\partial v}{\partial y}\right) - C_{Fx} \quad (4.53)$$

$$\frac{\partial v}{\partial t} + u\frac{\partial v}{\partial x} + v\frac{\partial v}{\partial y} = -\frac{1}{\kappa \rho C_\mu^2}\frac{\partial P}{\partial y} + \frac{1}{\rho C_R}\left(\frac{\partial}{\partial x}\mu\frac{\partial v}{\partial x} + \frac{4}{3}\frac{\partial}{\partial y}\mu\frac{\partial v}{\partial y} + \frac{\partial}{\partial x}\mu\frac{\partial u}{\partial y}\right.$$
$$\left. -\frac{2}{3}\frac{\partial}{\partial y}\mu\frac{\partial u}{\partial x}\right) - C_{Fx} \quad (4.54)$$

$$\frac{\partial \rho}{\partial t} + \frac{\partial \rho v}{\partial x} + \frac{\partial \rho v}{\partial y} = 0 \quad (4.55)$$

$$\frac{\partial \theta}{\partial t} + u \frac{\partial \theta}{\partial x} + v \frac{\partial \theta}{\partial y} = \frac{\kappa}{\rho C_v C_R \Pr} \left(\frac{\partial}{\partial x} \lambda \frac{\partial \theta}{\partial x} + \frac{\partial}{\partial y} \lambda \frac{\partial \theta}{\partial y} \right)$$

$$- \frac{P(\kappa-1)}{\rho C_v} \left(\frac{\partial u}{\partial x} + \frac{\partial v}{\partial y} \right) + \mu \frac{\kappa(\kappa-1)}{\rho C_v} C_\mu^2 \phi \quad (4.56)$$

$$\phi = 2 \left[\left(\frac{\partial u}{\partial x}\right)^2 + \left(\frac{\partial v}{\partial y}\right)^2 \right] + \left(\frac{\partial v}{\partial x} + \frac{\partial u}{\partial y} \right)^2 - \frac{2}{3} \left(\frac{\partial u}{\partial x} + \frac{\partial v}{\partial y} \right)^2$$

where ϕ is the dissipation function.

A perfect gas model is usually considered, with the thermodynamic equation of state $P = R\rho\theta$ (where R is the gas constant) which, in dimensionless form, is

$$P = \rho\theta \quad (4.57)$$

The thermal conductivity, dynamic viscosity, and specific heat capacity are assumed to depend only on temperature:

$$\lambda = \lambda(\theta) \quad \mu = \mu(\theta) \quad C_v = C_v(\theta) \quad (4.58)$$

We choose the quantities v_1, ρ_1, θ_1, L, $t_1 = L/v_1$, λ_1, μ_1, C_{v_1}, and C_{ρ_1} as the reference parameters for velocity, density, temperature, distance, time, thermal conductivity, viscosity, and heat capacity, respectively. The system (4.53)–(4.56) contains the following dimensionless groups based on these parameters:

$$C_R = \frac{v_1 L \rho_1}{\mu_1} \quad C_\mu = \frac{v_1}{\sqrt{\kappa_R \theta_1}} \quad \Pr = \frac{\mu_1 C_{\rho_1}}{\lambda_1}$$

$$C_{Fx} = \frac{F_x L}{v_1^2} \quad C_{Fy} = \frac{F_y L}{v_1^2} \quad \kappa = \frac{C_{\rho_1}}{C_{v_1}} \quad (4.59)$$

where F_x and F_y are the projections on the x and y axes of the external force vector. In the numerical investigation of particular physical problems, system (4.53)–(4.57) must include boundary and initial conditions based on the problem under consideration. The dimensionless groups (4.59), together with the dimensionless parameters obtained from the initial and boundary conditions and additional relations (4.58), represent a complete system of similarity parameters for the process under consideration. Which of the dimensionless groups should enter into the complete system depends on the particular problem.

CONVECTIVE HEAT TRANSFER

The finite-difference scheme developed by Polezhaev [20] was successfully used for a number of convection problems in viscous compressible gases and may be recommended here. The scheme is based on the fractional-step procedure and is of the explicit-implicit type. Inertia and viscous terms in the momentum equation are placed at the upper time level, and the pressure is left on the lower level. This gives a computational algorithm that is reducible to difference equations involving tridiagonal matrices and that may be solved quite economically by the factorization method.

Let us introduce a space-time difference grid whose points have coordinates (x_i, y_j, t^n), where $x_i = i\,\Delta x$ ($i = 0, 1, 2, \ldots, J$), $y_j = j\,\Delta y$ ($j = 0, 1, 2, \ldots, \mathcal{I}$), and $t^n = n\tau$ ($n = 0, 1, 2, \ldots$). Here, Δx and Δy are the grid steps for coordinates x and y, and τ is the time step.

The difference scheme approximating Eqs. (4.53)–(4.56) is

$$\frac{u_{i,j}^{n+1/2} - u_{i,j}^n}{0.5\tau} = \frac{4}{3\rho_{i,j}^n C_R}\tilde{\delta}_x^{\,2}(\mu u^{n+1/2}) - u_{i,j}^n \tilde{\delta}_x u^{n+1/2} + \frac{1}{\rho_{i,j}^n C_R}\tilde{\delta}_y^{\,2}(\mu u^n)$$
$$- v_{i,j}^n \tilde{\delta}_y u^n + f_u^{\,n+1/2} \quad (4.60)$$

$$\frac{v_{i,j}^{n+1/2} - v_{i,j}^n}{0.5\tau} = \frac{1}{\rho_{i,j}^n C_R}\tilde{\delta}_x^{\,2}(\mu v^{n+1/2}) - u_{i,j}^{n+1/2}\tilde{\delta}_x v^{n+1/2} + \frac{4}{3\rho_{i,j}^n C_R}\tilde{\delta}_y^{\,2}(\mu v^n)$$
$$- v_{i,j}^n \tilde{\delta}_y v^n + f_v^{\,n+1/2} \quad (4.61)$$

$$\frac{\rho_{i,j}^{n+1/2} - \rho_{i,j}^n}{0.5\tau} = -\frac{u_{i+1,j}^{n+1/2}\rho_{i+1,j}^{n+1/2} - u_{i-1,j}^{n+1/2}\rho_{i-1,j}^{n+1/2}}{2\Delta x}$$
$$- \frac{v_{i,j+1}^{n+1/2}\rho_{i,j+1}^n - v_{i,j-1}^{n+1/2}\rho_{i,j-1}^n}{2\Delta y} \quad (4.62)$$

$$\frac{\theta_{i,j}^{n+1/2} - \theta_{i,j}^n}{0.5\tau} = \frac{\kappa}{\rho_{i,j}^{n+1/2} C_{v_{i,j}} C_R \,\mathrm{Pr}}\tilde{\delta}_x^{\,2}(\lambda\theta^{n+1/2}) - u_{i,j}\tilde{\delta}_x \theta^{n+1/2}$$
$$+ \frac{\kappa}{\rho_{i,j}^{n+1/2} C_{v_{i,j}} C_R \,\mathrm{Pr}}\tilde{\delta}_y^{\,2}(\lambda\theta^n) - v_{i,j}^{n+1/2}\tilde{\delta}_y \theta^n + f_\theta^{\,n+1/2}$$
$$(4.63)$$

$$\frac{u_{i,j}^{n+1} - u_{i,j}^{n+1/2}}{0.5\tau} = \frac{4}{3\rho_{i,j}^{n+1/2} C_R}\tilde{\delta}_x^{\,2}(\mu u^{n+1/2}) - u_{i,j}^{n+1/2}\tilde{\delta}_x u^{n+1/2}$$
$$+ \frac{1}{\rho_{i,j}^{n+1/2} C_R}\tilde{\delta}_y^{\,2}(\mu u^{n+1}) - v_{i,j}^{n+1/2}\tilde{\delta}_y u^{n+1} + f_u^{\,n+1/2}$$
$$(4.64)$$

$$\frac{v_{i,j}^{n+1} - v_{i,j}^{n+1/2}}{0.5\tau} = \frac{1}{\rho_{i,j}^{n+1/2} C_R} \tilde{\delta}_x^2(\mu v^{n+1/2}) - u_{i,j}^{n+1} \tilde{\delta}_x v^{n+1/2}$$

$$+ \frac{4}{3\rho_{i,j}^{n+1/2} C_R} \tilde{\delta}_y^2(\mu v^{n+1}) - v_{i,j}^{n+1/2} \delta_y v^{n+1} + f_v^{n+1/2}$$

(4.65)

$$\frac{\rho_{i,j}^{n+1} - \rho_{i,j}^n}{0.5\tau} = -\frac{u_{i+1,j}^{n+1} \rho_{i+1,j}^{n+1/2} - u_{i-1,j}^{n+1} \rho_{i-1,j}^{n+1/2}}{2\Delta x}$$

$$- \frac{v_{i,j+1}^{n+1} \rho_{i,j+1}^{n+1} - v_{i,j-1}^{n+1} \rho_{i,j-1}^{n+1}}{2\Delta y} \quad (4.66)$$

$$\frac{\theta_{i,j}^{n+1} - \theta_{i,j}^{n+1/2}}{0.5\tau} = \frac{\kappa}{\rho_{i,j}^{n+1} C_{v_{i,j}} C_R \Pr} \tilde{\delta}_x^2(\lambda \theta^{n+1/2}) - u_{i,j}^{n+1} \tilde{\delta}_x \theta^{n+1/2}$$

$$+ \frac{\kappa}{\rho_{i,j}^{n+1} C_{v_{i,j}} C_R \Pr} \tilde{\delta}_y^2(\lambda \theta^{n+1}) - v_{i,j}^{n+1} \tilde{\delta}_y \theta^{n+1} + f_\theta^{n+1/2}$$

(4.67)

with the following notations for the difference operators:

$$\tilde{\delta}_x^2(\mu u) = \frac{1}{(\Delta x)^2}[\mu_{i+1/2,j} u_{i+1,j} - (\mu_{i+1/2,j} + \mu_{i-1/2,j})u_{i,j} + \mu_{i-1/2,j} u_{i-1,j}]$$

$$\tilde{\delta}_y^2(\mu u) = \frac{1}{(\Delta y)^2}[\mu_{i,j+1/2} u_{i,j+1} - (\mu_{i,j+1/2} + \mu_{i,j-1/2})u_{i,j} + \mu_{i,j-1/2} u_{i,j-1}]$$

$$\tilde{\delta}_x u = \frac{1}{2\Delta x}(u_{i+1,j} - u_{i-1,j})$$

$$\tilde{\delta}_y u = \frac{1}{2\Delta y}(u_{i,j+1} - u_{i,j-1})$$

The operators $\tilde{\delta}_x^2(\mu v)$, $\tilde{\delta}_y^2(\mu v)$, $\tilde{\delta}_x v$, $\tilde{\delta}_y v$, $\tilde{\delta}_x^2(\lambda \theta)$, $\tilde{\delta}_y^2(\lambda \theta)$, $\tilde{\delta}_x \theta$, and $\tilde{\delta}_y \theta$ are approximated similarly. In the right-hand sides,

$$f_u^{n+1/2} = -\frac{1}{\kappa \rho_{i,j}^n C_\mu^2} \frac{P_{i+1,j}^n - P_{i-1,j}^n}{2\Delta x} + \frac{1}{4\Delta x \Delta y \rho_{i,j}^n C_R}$$

$$\cdot \left[\left(\mu_{i,j+1} - \frac{2}{3}\mu_{i+1,j}\right) v_{i+1,j+1}^n + \left(\frac{2}{3}\mu_{i+1,j} - \mu_{i,j-1}\right) v_{i+1,j-1}^n \right.$$

$$\left. + \left(\frac{2}{3}\mu_{i-1,j} - \mu_{i,j+1}\right) v_{i-1,j+1}^n + \left(\mu_{i,j-1} - \frac{2}{3}\mu_{i-1,j}\right) v_{i-1,j-1}^n\right]$$

$$- C_{Fx}$$

$$f_v^{n+1/2} = -\frac{1}{\kappa \rho_{i,j}^n C_\mu^2} \frac{P_{i,j+1}^n - P_{i,j-1}^n}{2\Delta y} + \frac{1}{4 \Delta x \, \Delta y \, \rho_{i,j}^n C_R}$$

$$\cdot \left[\left(\mu_{i+1,j} - \frac{2}{3} \mu_{i,j+1} \right) u_{i+1,j+1}^{n+1/2} + \left(\frac{2}{3} \mu_{i,j+1} - \mu_{i-1,j} \right) u_{i-1,j+1}^{n+1/2} \right.$$

$$\left. + \left(\frac{2}{3} \mu_{i,j-1} - \mu_{i+1,j} \right) u_{i+1,j-1}^{n+1/2} + \left(\mu_{i-1,j} - \frac{2}{3} \mu_{i,j-1} \right) u_{i-1,j-1}^{n+1/2} \right]$$

$$- C_{Fy}$$

$$f_\theta^{n+1/2} = \frac{\mu_{i,j} \kappa (\kappa - 1) C_\mu^2}{\rho_{i,j}^{n+1/2} C_{v_{i,j}} C_R} \left\{ 2 \left[\left(\frac{u_{i+1,j} - u_{i-1,j}}{2\Delta x} \right)^2 + \left(\frac{v_{i,j+1} - v_{i,j-1}}{2\Delta y} \right)^2 \right] \right.$$

$$+ \left(\frac{u_{i,j+1} - u_{i,j-1}}{2\Delta y} + \frac{v_{i+1,j} - v_{i-1,j}}{2\Delta x} \right)^2 - \frac{2}{3} \left(\frac{u_{i+1,j} - u_{i-1,j}}{2\Delta x} \right.$$

$$\left. \left. + \frac{v_{i,j+1} - v_{i,j-1}}{2\Delta y} \right)^2 \right\}^{n+1/2} - \frac{\theta_{i,j}^n}{C_{v_{i,j}}} (\kappa - 1)$$

$$\cdot \left(\frac{u_{i+1,j} - u_{i-1,j}}{2\Delta x} + \frac{v_{i,j+1} - v_{i,j-1}}{2\Delta y} \right)^{n+1/2}$$

$$P_{i,j}^n = P(\rho_{i,j}^n, \theta_{i,j}^n) \qquad C_{v_{i,j}}^n = C_v(\theta_{i,j}^n) \qquad \lambda_{i,j}^n = \lambda(\theta_{i,j}^n)$$

$$\mu_{i,j} = \mu(\theta_{i,j}^n) \qquad \mu_{i+1/2,j} = \mu(\theta_{i+1/2,j}^n) \qquad \theta_{i\pm1/2,j}^n = \frac{1}{2}(\theta_{i\pm1,j}^n + \theta_{i,j}^n)$$

The system above approximates the initial system of differential equations (4.53)–(4.58) with the order $O(\tau + (\Delta x)^2 + (\Delta y)^2)$.

Difference equations (4.59)–(4.67) have a tridiagonal structure and may be solved by the factorization method. First, the values of the velocity components, density, and temperature are found at the intermediate time level $t^{n+1/2}$ by solving Eqs. (4.60)–(4.63) with the initial values taken at time t^n. Then, the above quantities are found for t^{n+1} from the solutions of equations (4.64)–(4.67).

Unlike the case in the numerical calculation of unsteady heat-convection problems, here the values of the stream functions are not calculated directly. With u, v, and ρ known, the stream functions ψ may be found with the equations

$$\frac{\partial \psi}{\partial x} = -\rho v \qquad \frac{\partial \psi}{\partial y} = \rho u \tag{4.68}$$

The following computational procedure may also be useful: Differentiating the first Eqs. (4.68) with respect to x and the second with respect to y and summing give the Poisson equation

$$\frac{\partial^2 \psi}{\partial x^2} + \frac{\partial^2 \psi}{\partial y^2} = \frac{\partial \rho u}{\partial y} - \frac{\partial \rho v}{\partial x} \tag{4.69}$$

The right-hand side is known and may be determined at any point of the grid (x_i, y_j). The solution of (4.69) for any suitable boundary conditions may be found by any of the methods described in Sec. 4.2.

In concluding this section, it should be pointed out that the density ρ at the boundary is not generally known from the problem conditions and must be found from the continuity equation (4.55) subject to the boundary conditions for velocity components u and v. Thus, if the no-slip condition $u = v = 0$ is prescribed at $x = 0$, then

$$\frac{\partial \rho}{\partial t} + \rho \frac{\partial u}{\partial x} = 0$$

Approximation of this expression by

$$\frac{\rho_{i,j}^{n+1/2} - \rho_{i,j}^n}{0.5\tau} = -\rho_{i,j}^n \frac{u_{1,j}^{n+1/2} - u_{0,j}^{n+1/2}}{h}$$

gives

$$\rho_{0,j}^{n+1/2} = \left(1 - \frac{\tau u_{1,j}^{n+1/2}}{2h}\right)\rho_{0,j}^n \qquad u_{0,j}^{n+1/2} = 0 \tag{4.70}$$

The boundary conditions for other quantities are approximated in the same way as in the difference schemes discussed for the (θ, ω, ψ) system.

For this difference scheme, the stability condition, found through many numerical experiments, is

$$\tau \leqslant \min\,(\Delta x, \Delta y)C_\mu \tag{4.71}$$

The difference scheme presented above is especially efficient for unsteady problems with low and moderate convection rates, since stability condition (4.71) is independent of u and v. For high convection rates, a steady solution is more difficult, since the difference scheme is nonmonotonic.

Chapter 5

Conclusion

5.1 GENERAL RECOMMENDATIONS FOR THE APPLICATION OF FINITE-DIFFERENCE TECHNIQUES TO HEAT-TRANSFER PROBLEMS

To use a finite-difference technique, one is faced with the problem of choosing the finite-difference procedure that will give the best approximate solution with the minimum amount of effort.

The choice of finite-difference scheme may be approached in different ways. If one is to solve a problem involving a small number of variations, then a very simple (e.g., explicit) scheme may be adopted. The programming then requires a minimum of effort. However, the computer time required to solve the problem may exceed that required by an economic difference scheme. But, because of the small number of variations to be computed, the excessive computer time is entirely compensated by the time saved in programming.

When one wants to investigate a certain class of problems systematically, or when the number of variations to be computed is large, then the use of an economical difference scheme is advisable. In spite of the more complicated structure of such schemes, resulting in longer programming times, their smaller computer-time requirements allow final results to be produced more quickly.

In heat-transfer problems, the determination of accurate heat-flux values is of great importance. For uniform heating and for moderate temperature gradients, most of the difference schemes considered give rather accurate values for the heat fluxes. However, for large temperature gradients and for nonuniform heating, conservative finite-difference schemes seem more useful, since in these schemes heat balance is satisfied more closely.

In going from a differential problem to its difference analog, the continuous set of arguments is replaced with a discrete set of points for which a numerical table is sought. It is very important that the characteristics of the original differential problem be retained at this point. The shape of the computational region should agree with the problem geometry. The shape of the mesh must therefore be chosen accordingly (square, rectangular, triangular, curved, etc.).

Regular grids with constant grid parameters are simplest. However, in the portions of the computational region in which solution gradients are especially large, it may be necessary to decrease the grid spacing to increase the accuracy of the numerical solution. In such cases, the substitution of new variables (e.g., the substitution $x' = \ln x$) may sometimes be recommended; a uniform grid would then be constructed with the new variables.

In general, the accuracy of the numerical solution and the required amount of computer time depend primarily on the selected grid parameters. In most cases, these parameters may be evaluated from numerical experiments.

The time step τ is generally dictated by the conditions of computational stability. It is very convenient to have the value of τ chosen automatically during computation. This is usually achieved by the proper construction of iteration cycles at every time level. The value of τ is normally based on the number of iterations required for the desired accuracy. The computation procedure is arranged as follows.

If the number of iterations Σ at the time level considered exceeds the prescribed number N_{max}, then the iteration procedure is stopped at the $(n + 1)$th time level, and the computation is repeated with step $\tau' = 1.3\tau$. If the desired accuracy requires a smaller number of iterations than N_{min},

CONCLUSION

then the computation is carried out with the time step $\tau' = 0.8\tau$. If $N_{min} \leqslant \Sigma \leqslant N_{max}$, a new time level is computed with $\tau' = 1.1\tau$. N_{min} and N_{max} depend on the particular problem; $N_{min} \approx 3\text{-}5$, and $N_{max} \approx 6\text{-}10$ are usually taken.

The optimal values for the space parameters of a grid depend on the required accuracy of the numerical solution. To find the optimal h, the most typical versions of the problem are successively computed with space steps of h, $h/2$, $h/4$, etc. The optimal h is then chosen by comparing the resultant solutions.

The smallest $h = h'$ for which the inequality

$$\frac{\max |\theta(h'/2) - \theta(h')|}{\max |\theta(h'/2)|} \leqslant \epsilon$$

is satisfied (where ϵ is the required relative accuracy) is considered as the optimal value.

In choosing the required accuracy ϵ, it should always be remembered that the accuracy of difference methods is usually not very high (within 1-20%). Accordingly, $\epsilon \approx 10^{-3}\text{-}10^{-4}$ is usually taken.

For steady-state problems, to save computer space, it may sometimes be useful to obtain the steady solution first from a rough grid, and then to decrease the grid parameters and extrapolate the resultant rough solution to additional grid points. The computation would then be continued until the steady solution was obtained on a fine grid.

If an approximate steady-state solution is required for a set of values of one parameter, then the solution obtained with a different value of the parameter may be recommended as the initial approximation; i.e., the computation should be carried out simultaneously for a set of parameter values.

References

1. Ryabenkii, V. S.; Filippov, A. E. "Ob ustoichivosti raznostnykh uravnenii" ("On the stability of difference equations"), p. 226, Moskva, Gostekhizdat, 1956.
2. Godunov, S. K.; Ryabenkii, V. S. "Raznostnye skhemy" ("Difference schemes"), p. 371, Moskva, Nauka, 1973.
3. Lax, P. D.; Richtmyer, R. D. Survey of the stability of linear finite difference equations, *Comm. Pure Appl. Math.*, no. 9, p. 267, 1956.
4. Neumann, J.; Richtmyer, R. D. A method for the numerical calculation of hydrodynamic shocks, *J. Appl. Phys.*, vol. 21, no. 3, pp. 232-243, 1950.
5. DuFort, E. C.; Frankel, S. P. Stability conditions in the numerical treatment of parabolic differential equations, *Math. Tables and Other Aids to Computation*, vol. 7, p. 135, 1953.
6. Samarsky, A. A. "Vvedenie v teoriyu raznostnykh skhem" ("Introduction to the difference scheme theory"), p. 496, Moskva, Nauka, 1971.
7. Samarsky, A. A.; Gulin, A. V. "Ustoichivost raznostnykh skhem" ("Stability of difference schemes"), p. 267, Moskva, Nauka, 1973.

8. Angel, E.; Bellman, R. "Dynamic programming and partial differential equations," p. 176, New York, Academic Press, 1972.
9. Young, D. Iterative methods for solving partial difference equations of elliptic type. *Trans. Am. Math. Soc.*, vol. 76, pp. 92–111, 1954.
10. Luikov, A. V. "Analytical heat-diffusion theory," p. 588, New York, Academic Press, 1968.
11. Sauliev, V. K. "Integrirovanie uravnenii parabolicheskogo tipa metodom setok" ("Integration of parabolic-type equations by the grid method"), p. 376, Moskva, Fizmatgiz, 1960.
12. Marchuk, G. I. "Metody vychislitel'noi matematiki" ("Computational mathematics methods"), p. 311, Novosibirsk, Nauka, 1973.
13. Boussinesque, J. "Theorie analytique de la chaleur," vol. 2, p. 172, Paris, Gautier-Villars, 1903.
14. Luikov, A. V.; Berkovsky, B. M. "Konveksiya i volny" ("Convection and waves"), p. 487, Moskva, Energiya, 1974.
15. Kuskova, T. V.; Chudov, L. A. O priblizhennykh granichnykh usloviyakh dlya vikhrya pri raschete techenii vyazkoi neszhimaemoi zhidkosti (On approximate boundary conditions for vortex in viscous incompressible flows), In "Vychislitel'nye metody i programmirovanie" ("Computational methods and programming"), no. 11, pp. 27–31, Moskva, MGU, 1968.
16. Berkovsky, B. M.; Nogotov, E. F. Teplovaya gravitatsionnays konveksiyapri nagreve sverkhu (Gravitational heat convection with heating from above), *Dokl. Akad. Nauk SSSR*, vol. 209, no. 1, p. 111, 1973.
17. Luikov, A. V.; Berkovsky, B. M.; Nogotov, E. F. Raznostnye metody issledovaniya zadach teplovoi gravitatsionnoi konvektsii (Difference methods for gravitational heat convection problems), In "Teplo. Massoperenos," vol. 8, p. 566, Minsk, Inst. of Heat and Mass Transfer of the Acad. of Sciences of the BSSR, 1972.
18. Berkovsky, B. M.; Nogotov, E. F. Fotoabsorbtsionnaya konvektsiya v polostyakh (Photoabsorptive convection in cavities), *Inzh. Fiz. Zh.*, vol. 19, no. 6, pp. 1012–1020, 1970.
19. Berkovsky, B. M.; Polevikov, V. K. Vliyanie chisla Prandtlya na strukturu i teploobmen pri estestvennoi konvektsii (Prandtl number effect on natural convection structure and heat transfer), *Inzh. Fiz. Zh.*, vol. 24, no. 5, pp. 842–849, 1973.
20. Polezhaev, V. I. Chislennoe issledovanie dvumernykh nestatsionarnykh uravnenii Navie-Stoksa szhimaemogo gaza v zamknutoi oblasti (Numerical investigation of two-dimensional unsteady Navier-Stokes equations for compressible gas in closed regions), *Izv. Akad. Nauk SSSR, Mekh. Zhidk. Gaza*, no. 2, pp. 103–111, 1967.
21. Polezhaev, V. I.; Griaznov, V. L. Metod raschota granichnuih uslovii

REFERENCES

dlia uravnenii Navie-Stoksa, *Dokl. Akad. Nauk SSSR*, vol. 219, no. 2, p. 301, 1974.

22. Yanenko, N. N. "Metod drobnuih shagov dlia reshenia mnogomernuih zadach matematischeskoi fiziki," Novosibirsk, Nauka, 1966.
23. Godunov, C. K.; Riabenikii, B. C. "Vvedenie v teoriu raznostunuih shem," p. 340, Moskva, Fizmatgiz, 1962.

Index

Accuracy:
  and convergence, 21
  order of, 16, 21, 107
Analytical expressions, 1
Approximation:
  of boundary conditions, 15, 37
  Boussinesque, 100, 102, 105, 125
  condition, 21
  of initial conditions, 18
  of Laplacian, 38, 40
  order of, 15, 21

Backward difference derivatives, 36
Balance method (*see* Integrointerpolation method)

Belotserkovsky, O. M., 2
Block-type iteration procedures, 56
Boundary conditions, 11, 59, 73–75, 78, 99–105, 116
  approximation of, 14, 15, 37, 69–76
  and convective transfer, 102
  of first kind, 70
  and Fourier transformation, 22
  of fourth kind, 71
  nonperiodic, 30
  and periodicity conditions, 22
  of second kind, 71
  and stability analysis, 30
  of third kind, 71
Boundary-value problems, 49

Boussinesque approximation, 100, 102, 105, 125

Cauchy problems, 22, 30, 41
Central-difference derivative, 37
Closed-form solutions, 1
Compressible media, 125
Computer use:
  approximation, 7, 86
  and grid choice, 35
  operating times, 56
  steady-state problems, 133
  storage, 51, 54, 79, 97, 125
Conditional probabilities, 3
Conjugate equations, 3
Conservation laws, 45
Conservative methods, 45, 47, 82, 132
Continuous-argument problems, 20
Continuum model, 2
Convective heat transfer:
  in compressible media, 4–5
  convection equations, 99–105
  high-rate processes, 116–122
  steady-state problems, 122–125
Convergence, 16–21
Correctness, of continuous-argument problems, 20
Crank-Nicholson scheme, 62, 63
Curvilinear grids, 36
Cylindrical geometries, 67–69

Damping, 19
Davydov, Y. M., 2
Difference formulas, for derivatives, 10–11
Difference operators, 12, 21
  linear, 20
  and point patterns, 36–40
Differencing grid, 2
Differential operators, 2, 8, 12, 36

Discontinuous-argument problems, 20
Dissipation function, 126
Dufort and Frankel, method of, 64

Eigenvalues:
  Neumann condition, 29
  spectral stability condition, 29
  of transition matrix, 28
Elimination method (*see* Gaussian elimination method)
Energy inequalities, 34
Error:
  damping, 19
  order of, 60
  and stability, 18
  truncation, 7
Eulerian grid, 2
Exact solutions, 16
Explicit schemes, 79, 95
Exponential error increase, 19

Factorization procedure (*see* Gaussian elimination method)
Finite-difference methods:
  definition, 11
  frozen coefficients, 30
  many-layered, 28
  substitutions, 36
Forward difference derivatives, 36
Fourier transformation, 24, 25, 30
  and stability, 22, 63
Fractional-step method, 4, 49, 56, 82, 84, 86, 106
Free convection, 99, 109
Frozen coefficients, 30
Function projection, 10

Gaussian elimination method, 33, 49, 51, 78, 90, 114, 127, 129

INDEX

Geometries, 132
Grashof number, 101, 109
Green formula, 112
Grids:
  boundaries, 36
  and convergence, 16
  curvilinear, 36
  functions, 9, 10, 12
  geometry of, 132
  for Laplacian, 39
  points, 2, 9, 10, 69
    interior, 40
    layers, 23
    patterns, 36–40
  regular, 35, 132
  steps, 9, 16
  uniform, 22

Harlow's large-particle technique, 2
Heat-conduction problems, 57–98
Heat-flux values, 132
Heuristics, 45
High-rate convective transfer, 123
Hydrodynamic equations, 2

Implicit difference schemes, 26, 61, 64, 79, 92, 95, 114, 120
Initial conditions, 11, 70
  approximation of, 18
Instability (*see* Stability)
Integrointerpolation scheme, 45–47, 76, 111

Jacobi iteration method, 54

Laplace equation, 38, 40, 53, 55
Large-particle method, 1, 2
Least-squares method, 48
Linear difference problems, 21

Linear operators, 23
Longitudinal-transverse schemes, 87, 90

Mass transfer, 1
Maximum principle, 30–34
Mesh (*see* Grid)
Metric, 34
Monte Carlo method, 1–2
Multidimensional problems, 2, 51, 84–85

Neumann condition, 29, 30
Nonlinear problems, 95
No-slip condition, 116, 130
Numerical experiments, 3
Nusselt number, 108

Operator form, 8, 12
Overrelaxation method, 54

Periodicity conditions, 22
Poisson equation, 40, 55, 102, 105, 107, 110, 122, 130
Polezhaev method, 127
Prandtl number, 101, 109
Predictor-corrector method, 89, 93
Probabilities, conditional, 3

Raleigh numbers, 5, 110, 124
Regular grids, definition, 35
Relaxation parameter, 54–55
Reynolds number, 101, 109

Saidel iteration method, 54, 55, 107
Samarsky method, 34, 118
Simpson's formula, 108

Smoothness conditions, 14
Spectral stability condition, 29
Spherical geometries, 67–69
Splitting-step method, 84
Stability, 16, 17, 32, 89, 93, 132
   conditions, 21, 23, 25, 43, 44, 51, 61, 78, 88–89, 114
   definition, 20
   and energy inequalities, 34
   and error, 18
   explicit schemes, 82
   Fourier transformation, 22–30, 63
   frozen coefficients, 30
   and grid choice, 35
   instability, 18, 19
   maximum principle, 30–34
   multidimensional analysis, 27
   nonlinear equations, 34
   Tom's condition, 103
   transition matrix, 27
Steady-state problems, 122, 133
Stefan-Boltzmann law, 71

Straight-line method, 1, 2
Substitutions, 2, 10

Taylor-series expansion, 13, 14, 37, 38, 43, 60, 102, 119
Thin rod equation, 8
Time levels, 19
Tom's condition, 103
Transfer effects, 2
Transition matrix, 27
Transition operators, 23, 25, 29
Truncation errors, 17

Variable coefficients, 76, 77
Variational methods, of Ritz and Galerkin, 48

Weighted difference schemes, 62
Woods' condition, 103

Yanenko, N. N., 2

TJ
260
N64
1978

FEB 12 1979